Maths for Biologists
Second Edition

A Pearson Custom Publication

Maths for Biologists
Second Edition

Compiled from:

Foundation Maths
Fourth Edition
by Anthony Croft and Robert Davison

Get Ready for Biology
Adapted from *Get Ready for A&P*
by Lori K. Garrett

And original material written by
Harriet Jones, Andrew Hemmings and Helen James

PEARSON
Custom Publishing

Pearson Education Limited
Edinburgh Gate
Harlow
Essex CM20 2JE

And associated companies throughout the worldw*Visit us on the World Wide Web at:*

www.pearsoned.co.uk

First published 2007

Taken from:

Foundation Maths
Fourth Edition
by Anthony Croft and Robert Davison
ISBN 978 0 13 197921 5

Get Ready for Biology
Adapted from *Get Ready for A&P*
by Lori K. Garrett
ISBN 978 0 321 50057 1

Original Material
Written by Harriet Jones, Andrew Hemmings and Helen James

ISBN 978 1 84776 241 2

Printed in the United Kingdom by Henry Ling Limited, at the Dorset Press, Dorchester, DT1 1HD

Contents

Foreword

As a scientist it is important to understand how mathematics relates to the study of scientific subjects and the scientific method. One of the most important historical advances in science was the use of mathematics and measurement to give precise accounts of observation and experience. This happened around the fifteenth century. Combining mathematics and what, at this time, was natural philosophy was arguably the birth of the modern scientific method. There was a gradual change at this time from a situation where mathematics was admired for its own sake, to one where mathematics was tool that revealed how things must be; if the calculations worked the theory could be supported more strongly. Copernicus used mathematics to support his argument that it was the Earth that moved around the sun, rather than the other way around. His critics believed in philosophical and physical justifications, but with Copernicus his evidence was purely mathematical; this was a major change in the approach to science.

From the use of mathematics in astronomy came its use in terrestrial subjects such as mechanics, navigation and warfare. Because of this, mathematics began to gain higher social, as well as scientific, status. A mathematical approach to these subjects was gradually replacing the more descriptive Aristotelian natural philosophy. In other words, without mathematics it was harder to support a theory. Galileo was a particularly important historical figure in that he used mathematics to understand the world around him, rejecting much of what Aristotle had taught. For example, Aristotle had taught that force must be continuously be applied to an object to keep it in motion. Through experimentation and mathematics Galileo was able to dispute this. Newton took many of these early ideas forward and, using mathematics, demonstrated many laws of science that lead to a much greater understanding of the world. The work of Newton was acceptable in its time because of the work of his predecessors who had combined mathematics with natural philosophy and made the study of mathematics intellectually and sociably acceptable.

Following the application of mathematics to science came the application of the experiment, replacing the study of experiences; the replacement of the qualitative with the quantitative. From this came the need to measure, not just for mathematicians, but for natural philosophers too. This, around the time of the seventeenth century, lead to the development of instruments for calculating and for measuring. Science and mathematics became inseparable. Some of the most impressive and historically important experiments, such as those carried out in the nineteenth century by Mendel on peas, were completely based on mathematics and a statistical treatment of data. As a scientist, therefore, it is important to develop an appreciation of the power of mathematics and how crucial it is to a thorough and deep understanding of science.

Front cover picture is of a swan on the UEA Broad taken by David Boulton in February 2008 (David Boulton Photography, www.davidboulton.co.uk).

About this book

This book is designed to help students with a GCSE in mathematics, or equivalent, progress to understand the mathematics required in a biological science degree.

Some chapters in this book are purely mathematical and others look at biological applications of the mathematics. A general ability to manipulate numbers and think mathematically will help you understand many biological issues, and help you interpret data you produce yourself. There are problems for you to try in each chapter that test your understanding of each subject covered.

The first chapter in this book allows you to test yourself, to make sure you reach the absolute minimum standard required to start a biology degree programme. It may be that you haven't done any mathematics for a long time. If so it will serve as a revision in the basics for you. It is, though, a very basic level and if you struggle with any of it you need to work especially hard and seek extra help.

This book has been compiled with the kind help and support of many Faculty at the School of Biological Sciences at the University of East Anglia. Specifically thanks are due to Dr Helen James, Dr Kay Yeoman, Prof Tony Davy and Dr Richard Bowater for advice and proof reading the material written by Dr Harriet Jones. Dr Robert Jenkins, from the Dean of Students' Office, has provided invaluable support and advice in the writing of material additional to that supplied from current text books. Dr Andrew Hemmings very kindly supplied the material for the chapter on calculus.

Foundation Maths

Visit the *Foundation Maths, fourth edition* Companion Website at **www.pearsoned.co.uk/croft** to find valuable **student** learning material including:

- Student Support Pack. Contains one-page summaries of key topics complete with examples, exercises and answers
- Extra end-of-chapter questions with answers

Chapter 1
Basic Maths Revision

Crunching the Numbers

When you complete this chapter, you should be able to:

- Solve math problems involving multiple operations, fractions, decimals, and percents.
- Calculate the mathematical mean.
- Work with exponents, numbers in scientific notation, ratios, and proportions.

Your Starting Point

Answer the following questions to assess your math skills.

1. 2/3 × 3/4 = ____________
2. Express 50% as a fraction: ____________
   As a decimal: ____________
3. What is the *mean* of 27, 33, and 36? ____________
4. In scientific notation, 1000 = 10 ____
5. 30% of 200 = ____________
6. Which is longer, 1 yard or 1 meter? ____________
7. In the metric system, the basic unit of volume is the ____________.
8. Assume you take 15 quizzes by the end of the semester and get 9 A's and 6 B's. Express this as a ratio. ____________
9. Is a triple beam balance used to measure volume, mass, or distance?
   ____
10. On a graph, the vertical line is the ____________ axis.

How Much Wood Would a Woodchuck Chuck? **Math in Science**

You probably remember doing story problems when learning math in your younger years. Those problems helped you see how math can be used. Many students are surprised to learn that they have to use math in biology. But you must remember that science—all science—deals with that which is testable. A scientific test, as you know, is called an experiment. Results collected from experiments are called **data** and, more often than not, the data are numbers. When you try to make sense of the data, you are working with numbers, and that means math.

Answers: 1. 1/2. 2. 1/2, 0.5, 3. 32. 4. 10^3. 5. 60. 6. 1 meter. 7. liter. 8. 9:6 = 3:2. 9. mass. 10. y.

You will likely do some experiments in lab and then analyze the data. These experiments may involve the study of physiology, the internal processes of living organisms, as a part of biology. For example, you may investigate one of the many aspects of body function that have "normal" conditions; these are often expressed in numerical values. For example, normal human body temperature is 98.6°F, normal blood pressure is 120/80, and normal pulse is around 70 to 80 beats per minute. In addition, you will work with chemical solutions and you will need to understand their concentrations. You will also measure in metric units, refer to percentages and ratios, and interpret graphs and charts.

Many students entering this class may only need a brief reminder of what they learned before, whereas others may need to learn it again. Regardless of your math history, a quick refresher will help you better understand the numbers.

From the Beginning: **Basic Math Operations**

You might need to do some complicated computations in class, so it is good to remember the basic rules. Let's zip through multiplication and division for a quick refresher.

MULTIPLICATION Multiplication problems can be done in any order: $3 \times 4 \times 2 = 24$, or $2 \times 3 \times 4 = 24$. The answer to the equation is called the **product**. Recall that multiplying any number by 1 does not change the number, while multiplying any number or numbers by 0 gives you 0.

In more complicated equations, an operation may be set off in parentheses or brackets. If a number appears immediately to the left of a parenthesis or bracket, multiply by that number even though there is no multiplication sign:

$$3(6 - 2) = 3 \times (6 - 2) = 3 \times 4 = 12$$

Multiplication problems are sometimes represented with **exponents**. Consider 2^4, which is read as "two to the fourth power," and 10^3, which is read as "ten to the third power." These examples are really just a shorthand way of expressing these multiplication problems:

$$2^4 = 2 \times 2 \times 2 \times 2 = 16 \qquad 10^3 = 10 \times 10 \times 10 = 1000$$

DIVISION Like subtraction, division problems must be done from left to right. Consider this example:

$$10 \div 5 = 2,$$
$$\text{but } 5 \div 10 = 1/2$$

In division, the number being divided is called the **dividend**, the number by which it is divided is the **divisor**, and the total is the **quotient**. In our first example, 10 is the dividend, 5 is the divisor, and 2 is the quotient.

MULTIPLE OPERATIONS Let's try some more complicated problems that involve more than one mathematical operation. Parentheses or brackets are often used in equations when there are multiple operations. Think of them as directors telling you how to proceed. Always do operations within these structures first. Let's see why this matters.

$$8 - (2 \times 3) = 8 - (6) = 2,$$
$$\text{but } 8 - 2 \times 3\text{, done in that order,} = 6 \times 3 = 18$$

OOPS!

Always approach an equation by first doing any operations within parentheses.

Here are some simple rules to help ensure that you perform mathematical operations in the correct order:

1. First, do all operations inside the parentheses or brackets.
2. Next, multiply out any exponents.
3. Then do all multiplication and division equations, moving from left to right.
4. Finally, do all addition and subtraction problems, again from left to right.

TIME TO TRY

Try the following problems.

1. $4 \times (9 - 6) + 10 =$ _______

2. $3^3 \div 9 - 4 + 5 =$ _______

3. $6^2 - 2(5 - 2) + 4 - 2 =$ _______

Let's see how you did.

Problem #1: The correct answer is 22. First, do what is in the parentheses (rule 1): $(9 - 6) = 3$, so the problem becomes $4 \times (3) + 10$. Next, do the multiplication (rule 3): $4 \times 3 = 12$, so the problem becomes $12 + 10$. Finally, do the addition (rule 4): $12 + 10 = 22$.

Problem #2: The correct answer is 4. There are no parentheses, so you start with the exponent (rule 2): $3^3 = 3 \times 3 \times 3 = 27$, and the problem becomes $27 \div 9 - 4 + 5$. Next, do the division (rule 3): $27 \div 9 = 3$, so the problem becomes $3 - 4 + 5$. Finally, do the addition and subtraction from left to right (rule 4), and you get $3 - 4 = -1$, then $-1 + 5 = 4$.

Problem #3: The correct answer is 32. Start in the parentheses (rule 1): $(5 - 2) = 3$, so the problem becomes $6^2 - 2(3) + 4 - 2$. Next, take care of the exponent (rule 2): $6^2 = 6 \times 6 = 36$, so the problem becomes $36 - 2(3) + 4 - 2$. Now, do the multiplication and division from left to right (rule 3): $2(3) = 2 \times 3 = 6$ and $4 - 2 = 2$, so the problem becomes $36 - 6 + 2$. Finally, do the addition and subtraction from left to right (rule 4): $36 - 6 = 30$, then $30 + 2 = 32$.

As you see, some mathematical equations can be long and somewhat complicated, but if you keep the basic rules in mind and tackle them step by step, they become quite manageable.

What Do You **Mean** You Are Normal?

Do you know what "normal" body temperature is? Sure you do—98.6°F. You have likely known that since you were a small child. But what is YOUR normal temperature? What if yours is rarely 98.6°? "Normal" blood pressure is 120/80. What if yours runs lower than that? If you think this makes you abnormal, reconsider what the term "normal" really means.

In biology, the term **normal** means **average**. And in math, another term for average is **mean**. We refer to many normal values—temperature, blood pressure, pulse, respiratory rate . . . the list goes on and on. When you see these, realize that they are average values and an individual person may have a different normal value—what is normal for him or her may not be average for the whole population. To better understand this idea of normal, you need to know how to calculate the average, or mean, value.

Let's say you're doing a lab on the cardiovascular system (heart and blood vessels) and you're measuring pulse rate. You are instructed to do three trials and then calculate the mean pulse rate. Your three trials give you the following data:

Trial 1: 72 beats per minute
Trial 2: 74 beats per minute
Trial 3: 79 beats per minute

To find the mean of a group of numbers, simply add them all together then divide the total by how many numbers you added. For your data, you would add the three pulse rates, then divide by 3:

$$72 + 74 + 79 = 225 \qquad 225 \div 3 = 75 \text{ beats per minute}$$

Did you notice that the mean is not one of the original numbers? It does not have to be. It is the average of all three numbers.

Here is a tip to help you with means and with most math problems—learn to predict your results! The mean of a group of numbers will be somewhere between the highest and the lowest of the numbers you are averaging. If your value does not fall in that range, check to see if you made an error. Common errors are missing a number during the addition or dividing by the wrong number. If you estimate your result first, you can more easily recognize errors if they occur.

TIME TO TRY

Calculate the mean of these body temperatures:

97.4°F, 98.0°F, 99.2°F, and 99.8°F

1. What is the mean? _______ °F
2. What does this mean *mean*? ______________________________

____

If you did this correctly, you should have gotten a mean of 98.6°F, even though that was not one of the original temperatures listed. So, as stated earlier, the physiological "normal" value is a mean, and individuals' normal temperatures will vary around that mean.

WHY SHOULD I CARE?

All science is based on data and experimental trials. There is a certain amount of error possible with each trial. Consider the pulse values we used as examples. Three trials gave us three results. Using the mean helps to minimize the error from individual trials.

✔ **QUICK CHECK**

What is meant by saying that normal human heart rate is 80 beats per minute?

____

Answer: It means the average, or mean, heart rate for humans is 80 beats per minute.

Meet My Better Half: **Fractions, Decimals, and Percents**

Working with whole numbers is rather easy and is second nature to most of us. Some of us may need to brush up on fractions, however. Related to fractions are two other ways of expressing values: decimals and percents. In biology, all three of these will be used. For example, the micrometer (μm), a common unit for measuring the size of microscopic structures, is a tiny fraction of the more familiar millimeter—a micrometer is 1/1000 of a millimeter, to be exact. As you just saw, normal body temperature is reported as a decimal (98.6). Finally, about 60% of the body of an average adult male is water. You will discover that many biological values are reported in any of these formats, so you want to be comfortable with their use.

FRACTIONS

Fractions are written as *a/b*, in which *a* and *b* are both whole numbers and *b* is not 0. The first (top) number is called the **numerator**, and the one on the bottom is the **denominator**. A fraction represents parts of some whole group (**Figure 1.1**). For example, 3/4 represents 3 equal parts out of 4 equal parts, where the 4 equal parts make up the whole (Figure 2.1a). Whole numbers can be represented as fractions as well.

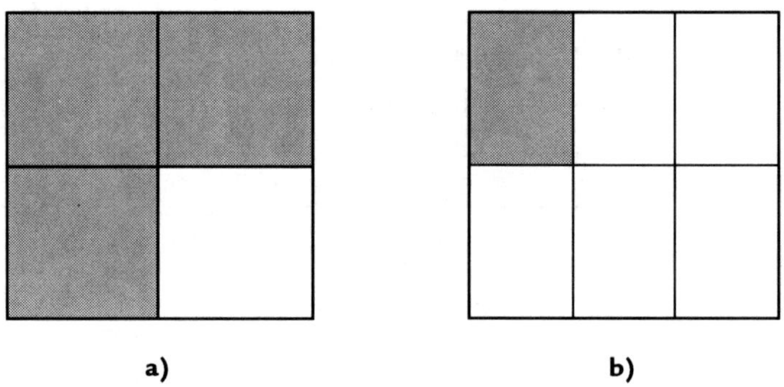

FIGURE 1.1 Fractions represent some part of a whole. In each of these squares, the shaded area represents the fraction, as follows: **a)** This square is divided into four equal parts, and 3 of the four are shaded = 3/4. **b)** This square is divided into six equal parts, and 1 of the 6 is shaded = 1/6.

The whole number simply becomes the numerator, and the denominator is 1, so 3 = 3/1.

✔ **QUICK CHECK**

For 5/8, what is the numerator? _______ The denominator? _______
Express 6 as a fraction: _______

Answers: 5 is the numerator, 8 is the denominator, and 6 as a fraction is 6/1.

REDUCING FRACTIONS **Equivalent fractions** have the same value even though they appear to be different. Consider the following fractions:

1/3 2/6 4/12 7/21

All of these numbers have the same value: 1/3. To see this, you need to **reduce** the other fractions. This is done by finding the **greatest common factor (GCF)** for each fraction. The greatest common factor is the largest whole number that can be divided into both the numerator and the denominator. Consider 2/6. Both the numerator (2) and the denominator (6) are divisible by 2, which is the greatest common factor. If you do the division, you see that 2 ÷ 2 = 1 and 6 ÷ 2 = 3, so 2/6 becomes 1/3.

TIME TO TRY

Look at the other fractions we listed: 4/12 and 7/21.

What is the greatest common factor for 4/12? _______

Divide the numerator by that factor: _______
Divide the denominator by that factor: _______

What is the reduced fraction? _______

What is the greatest common factor for 7/21? _______

Divide the numerator by that factor: _______
Divide the denominator by that factor: _______

What is the reduced fraction? _______

How did you do with Time to Try? You should have found that the greatest common factor for 4/12 is 4, so $4 \div 4 = 1$ and $12 \div 4 = 3$. Thus, the fraction 4/12 reduces to 1/3. Similarly, 7/21 has a greatest common factor of 7, and $7 \div 7 = 1$, and $21 \div 7 = 3$ so, again, 7/21 reduces to 1/3. After using a number to reduce the fraction, check your result to see if it is in its simplest form or if it can be reduced further.

✔ **QUICK CHECK**

What is the most reduced form of each of the following fractions: 60/90, 25/100, and 18/54?

Answer: 60/90 has a GCF of 30 and reduces to 2/3; 25/100 has a GCF of 25 and reduces to 1/4; 18/54 has a GCF of 18 and reduces to 1/3.

MULTIPLYING AND DIVIDING FRACTIONS When doing mathematical operations with fractions, the rules are different for multiplication and division than they are for addition and subtraction. For multiplication, you simply multiply the numerators in one step, then multiply the denominators. Consider $2/3 \times 3/4$. The numerators are 2 and 3. Multiply them to get 6, and that goes on top. Next, multiply the two denominators, 3×4, to get 12. So the product is 6/12, which reduces to 1/2:

$$\frac{2}{3} \times \frac{3}{4} = \frac{6}{12} = \frac{1}{2}$$

Let's try another: $1/3 \times 2/5 \times 3/4 =$ __________

First, multiply all the numerators ($1 \times 2 \times 3 = 6$), then multiply all the denominators ($3 \times 5 \times 4 = 60$) and you get 6/60, which reduces to 1/10.

To multiply fractions, first multiply all the numerators, then multiply all the denominators. Reduce the result as needed.

Dividing fractions may seem difficult at first, but a simple trick actually makes it easy! These problems may be written two different ways:

$$\frac{4/5}{2/3} \text{ or } 4/5 \div 2/3$$

Solving them is easy. First, invert (flip) the second fraction, which is the divisor: 2/3 becomes 3/2. Then you simply multiply the two fractions:

$$4/5 \div 2/3 = 4/5 \times 3/2 = 12/10$$

Now, 12/10 can be reduced to 6/5. Here the numerator is larger than the denominator, which means this fraction is greater than 1. Usually when this happens, it is best to express the answer as a mixed number—one combining both whole numbers and fractions. To do this, first reduce the fraction: $12/10 = 6/5$. Then realize that $6/5 = 5/5 + 1/5$. Since 5/5 equals 1, the mixed number would be 1-1/5 (read as 1 and 1/5).

To divide one fraction by another, first invert the second fraction to turn it into a multiplication problem. Next, multiply the numerators, then multiply the denominators. Finally, reduce the result.

ADDING AND SUBTRACTING FRACTIONS To add or subtract fractions, they must first be in the same format. You might think you can just add the numerators and denominators, but that won't work. By that method, $1/2 + 1/4$ would equal 2/6, which reduces to 1/3. But that is smaller than 1/2, one of the numbers we added! This doesn't make sense. (See why it helps to predict your results?) Instead, you must first put the fractions into common terms. They must have the same denominator, called a **common denominator**.

To get the common denominator, you need to know the **least common multiple (LCM)**. This is the smallest number that can be divided by both the numerator and denominator. In our example of $1/2 + 1/4$, 4 is the least common multiple, so we want both fractions to have 4 as their denominator. Recall that any number multiplied by 1 does not change. To convert 1/2 into fourths, we multiply it by 2/2 (=1). Thus, $1/2 \times 2/2$ becomes 2/4. Once the fractions have a common denominator, we simply add the numerators only:

$$2/4 + 1/4 = 3/4$$

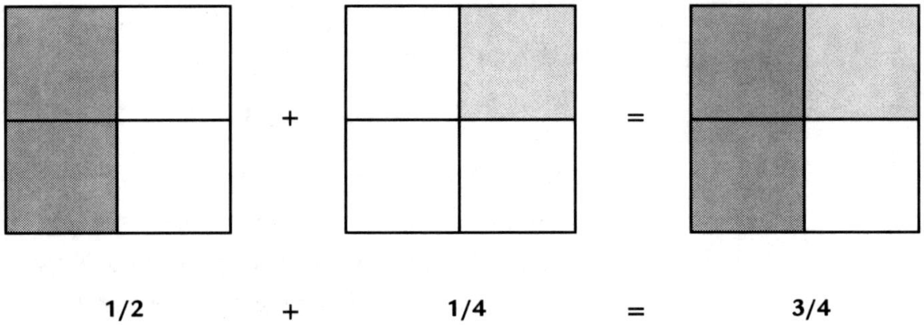

FIGURE 1.2 Visual confirmation that adding 1/2 + 1/4 = 3/4.

Figure 1.2 illustrates this for you.

Subtracting fractions also requires a common denominator. Once you have the denominator, simply subtract the second numerator from the first numerator. Let's try one: 1/3 − 1/4. The smallest common denominator for these two fractions is 12, so they both need to be converted, as follows:

$$1/3 \times 4/4 = 4/12$$
$$1/4 \times 3/3 = 3/12$$

Now line up your fractions in the correct order from left to right, then subtract the second numerator from the first:

$$4/12 - 3/12 = 1/12$$

To add or subtract fractions, use a common denominator to put the fractions in a common form, then add or subtract the numerators only. Remember to always subtract from left to right.

✔ **QUICK CHECK**

Solve these problems:

1. 3/5 + 1/4 + 1/10 = ________
2. 9/16 − 3/8 = ________

Answers: 1. The smallest common denominator is 20, so 3/5 × 4/4 = **12/20**, 1/4 × 5/5 = **5/20**, and 1/10 × 2/2 = **2/20**. Then add the numerators: 12 + 5 + 2 = **19/20**. 2. The smallest common denominator is 16, so 3/8 × 2/2 = **6/16**. Then 9/16 − 6/16 = **3/16**.

DECIMALS

Decimals are common in science, and biology is no exception. All decimals are based on 10 in a very specific way—each place in the number represents a multiple of 10. The value *increases* 10 times for each space that you move to the left of the decimal, and it *decreases* 10 times for each space to the right.

Let's consider this number: **12345.6789**. You know how to read the numerals to the left of the decimal. They make 12,345. Moving from the decimal to the left, you can see how the spaces represent, in order, 1's, 10's, 100's, 1000's, and 10,000's (**Figure 1.3**). Our number represents 10,000 + 2000 + 300 + 40 + 5. The spaces to the right of the decimal represent fractions: tenths, hundredths, thousandths, and so on. Thus in our number, as you go to the right of the decimal, the numerals represent the fractions 6/10, 7/100, 8/1000, and 9/10,000.

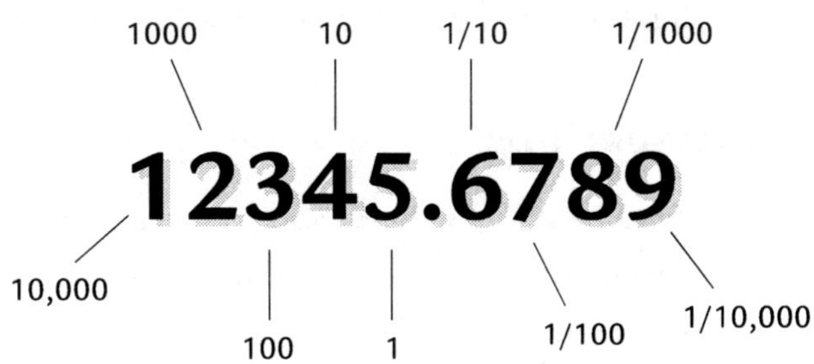

FIGURE 1.3 Each space around a decimal point reflects a change by a factor of 10.

CONVERTING DECIMALS As with fractions, decimals allow you to express a number more precisely than you can with whole numbers. In fact, you can think of fractions as division problems that give you a decimal value, so fractions can be converted to decimals. For example, 1/2 is $1 \div 2$, and that equals the decimal 0.5. Note the 0 here—it is used as a place holder so that you know where the decimal belongs. If you use a calculator to find $1 \div 3$, the answer will be 0.333333 . . . . This is known as a **repeating decimal**. When you have a repeating decimal, one option is to round off. If the last number is less than 5, round down; if it is 5 or higher, round up. For example, 0.33333 . . . would round down to 0.33, and 0.66666 . . . would round up to 0.67.

Decimals can also be converted into fractions. The value 0.25 represents $2/10 + 5/100$. When you add them, remember to first get a common denominator. You get $20/100 + 5/100 = 25/100$, and that reduces to 1/4.

ADDING AND SUBTRACTING DECIMALS Adding and subtracting decimals is easy as long as you line them up correctly. Consider this example: $1.287 + 24.32$. First, be sure the two numbers have the same number of spaces after the decimal. The first number has three places, but the second only has two. But, you can always add zeros to the end of a decimal number, $24.32 = 24.320$. Note that the last 0 means there are 0/1000, which is correct and does not change the value. Finally, it is easiest to add these numbers if you line them up vertically, always being sure the decimal points are aligned:

$$\begin{array}{r} 1.287 \\ +\ 24.320 \\ \hline =\ 25.607 \end{array}$$

Subtraction is done the same way.

PICTURE THIS

Assume that you have worked three extra jobs for pocket cash this week. From them, you earned \$33.70, \$45.28, and \$21.02. How much extra money did you earn? ______________

With this money you buy a pizza for \$8.99, soft drinks for \$1.49, gas for \$20.00, and a new CD for \$19.95. How much money do you have left? ____________

Congratulations, you just added and subtracted decimals, as you do on a regular basis in daily life! You should see that you earned \$100 and have \$49.57 left.

When adding or subtracting decimals, always align the decimal point in the two numbers before doing the operation.

MULTIPLYING AND DIVIDING DECIMALS Multiplication and division of decimals is a bit trickier because you must keep track of how many decimal places you should have at the end. Let's try an easy one: 0.5×0.3. First, multiply the numbers as if they are whole numbers: $5 \times 3 = 15$. Now, add the number of decimal places you started with. Both numbers you multiplied originally had one decimal place, so that adds up to two. Realize that your answer of 15 is 15.0, so you know where the decimal begins. Now you have to move the decimal. The numbers you started with had a total of two digits after the decimal, so you must move the decimal point left by two places, giving you 0.15. Here is a way to double check that. If the original numbers were fractions, they would be 3/10 and 5/10. Recall how to multiply fractions—you multiply the numerators, then multiply the denominators:

$$3/10 \times 5/10 = 15/100 = 0.15$$

What if the numbers had been 0.03×0.5? Although you still get 15, now you need to move the decimal three places to the left, but there are only two. You simply add zeros to the left until you have the correct number of decimal places, in this case giving you 0.015.

Division with decimals is just like ordinary division, except we keep going until we either finish or reach a predetermined stopping point. You usually do not go beyond the number of decimal points your original numbers contain, so if they had a total of two, you would likely stop at two and round off beyond that. Let's look at an example: 2.8 ÷ 7. Put the decimal point in the answer line exactly above its position in the dividend (2.8), then simply do the division:

$$\begin{array}{r} 0.4 \\ 7\,\overline{)\,2.8} \\ -2\,8 \\ \hline 0 \end{array}$$

How do you divide when both numbers are decimals? Let's divide 2.1 by 0.7. All you have to do is move the decimal of the divisor until you have a whole number, then move the decimal of the dividend by the same number of spaces in the same direction. If you try 1.68 (dividend) ÷ 0.3 (divisor), you move the decimal in 0.3 one spot to the right to get 3, then you must also move the decimal in 1.68 one spot to the right, getting 16.8, so the problem becomes 16.8 ÷ 3. Do long division to get the result:

$$\begin{array}{r} 5.6 \\ 3\,\overline{)\,16.8} \\ -15 \\ \hline 18 \\ -18 \\ \hline 0 \end{array}$$

You can always multiply back to double check your result:

$$0.3 \times 5.6 = 1.68$$

Moving the decimal may seem confusing, but there are some easy shortcuts to remember.

- Moving the decimal to the *right* one space is the same as multiplying by 10; two spaces multiplies by 100; and so on, so the numbers get *bigger*.
- Moving the decimal to the *left* means you are dividing by 10 for each space moved, and the number always gets *smaller*.

Moving the decimal to the right is the same as multiplying by 10 for each space moved. Moving it to the left is the same as dividing by 10 for each space moved.

QUICK CHECK

Solve these problems:

1. $1.27 + 3.6 =$ __________
2. $14.87 - 3.2 =$ __________
3. $2.4 \times 1.2 =$ __________
4. $8.4 \div 0.2 =$ __________

Answers: 1. $1.27 + 3.60 = 4.87$. 2. $14.87 - 3.20 = 11.67$. 3. $2.4 \times 1.2 = 2.88$. 4. Move the decimal in the divisor and dividend both one space to the right, so it becomes $84 \div 2 = 42$.

PERCENTS

As you learn biology, you will encounter many values that are stated as percents. Percents are based on 100, where 100% is the total. For that reason, when working with percents, always be sure they add up to 100% and no more than that.

Percents are easy to work with. They are essentially fractions expressed as hundredths. For example, 25% is the same as 25/100, which can be further reduced to 1/4. And because fractions can be expressed as decimals, so can percents. You simply put the decimal two places to the left of the percent. So, 25% becomes 0.25 and 7% becomes 0.07. Percents, decimals, and fractions are all interchangeable, but when doing math operations, percents should be converted into either decimals or fractions. You cannot do math operations with mixed expressions—they must all be whole numbers, or fractions, or decimals. **Table 1.1** explains the relationship between these expressions.

TABLE 1.1 The relationship between percents, decimals, and fractions.

Percent	Decimal	Fraction in hundredths	Reduced fraction
10%	0.10	10/100	1/10
25%	0.25	25/100	1/4
40%	0.40	40/100	2/5
50%	0.50	50/100	1/2
75%	0.75	75/100	3/4
100%	1.00	100/100	1

TIME TO TRY

Can you supply the missing information in this table?

Percent	Decimal	Fraction
36%	______	______
______	0.42	______
______	______	80/100 = 4/5

Answers: 36% = 0.36 = 36/100 = 9/25; 42% = 0.42 = 42/100 = 21/50; 80% = 0.8 = 80/100 = 4/5.

Can You Feel the Power? **Understanding Exponents**

We briefly discussed exponents earlier, when we stated that a number written with an exponent is basically a multiplication problem: $2^3 = 2 \times 2 \times 2 = 8$. In science, very large and very small numbers are often written in a special format that uses exponents based on powers of 10. This format is called **scientific notation**. Let's consider the number 200 to see how this format is used. To write a number in scientific notation, first place the decimal immediately after the first digit, and then drop the zeroes. This number is called the **coefficient**. In this example, the coefficient is 2. Next, count how many spaces you moved the decimal—two places to the left. Each of those spaces represents a power of 10, so two places means $10 \times 10 = 100$. In scientific notation, we would write 200 as 2×10^2. As you can see, this is $2 \times 10 \times 10 = 200$. Remember, the first number in scientific notation must be greater than 1 but less than 10. If a number is less than 1, the exponent is a negative power of 10. For example, 0.0004 would be 4×10^{-4} because the decimal was moved four spaces to the right. You will rarely need to do math operations with scientific notation, so we will skip those. You should, however, understand scientific notation so that you can understand some of the measurements you will read about—such as cell sizes, which are often measured in micrometers (10^{-6} meter). **Table 1.2** lists some common exponents.

TIME TO TRY

Express the two numbers below in scientific notation.

24,000,000 = ________________

0.003 = ________________

If you did this correctly, you got 2.4×10^7 and 3×10^{-3}.

TABLE 1.2 The values of some common exponents used in scientific notation.

Exponent	Value	Term
10^9	1,000,000,000	billions
10^6	1,000,000	millions
10^3	1000	thousands
10^2	100	hundreds
10^1	10	tens
10^0	1	ones
10^{-1}	1/10	tenths
10^{-2}	1/100	hundredths
10^{-3}	1/1000	thousandths
10^{-6}	1/1,000,000	millionths
10^{-9}	1/1,000,000,000	billionths

My Cell's Bigger than Your Cell! **Ratios and Proportions**

- Sodium and potassium move across cell membranes in a 3:2 relationship.
- In the United States, the ratio of males to females at birth is about 105:100.
- The ratio of males to females declines steadily until, after age 85, it is only 40.7:100.

HUH? Welcome to the comparatively interesting realm of ratios. A **ratio** expresses a relationship between two or more numbers—it is a way to compare them. Ratios can be expressed using a colon between the numbers (as above), as a fraction, or by using the word "to." For example, carbohydrates contain hydrogen and oxygen in a 2 to 1 ratio, meaning there are twice as many hydrogen atoms in carbohydrates as there are oxygen atoms.

Ratios are used for comparison, and they can also be expressed as fractions. For example, a ratio of 1:2 means the same as 1/2. Look at that carefully, though. Let's say the ratio of men to women in your biology class is 1:2. We're not saying that half of the class are men, we are saying there are half as many men as women.

If ratios can be expressed as fractions, they can also be expressed as decimals and percents. Because they can be written as fractions, they can also be reduced like fractions. For example, a ratio of 4:6 is the same as 4/6, which is the same as 2/3. When working with ratios, it is critical to write them in the correct order. If a biology class has 10 males and 20 females, the ratio of males to females is 10:20. If we write it as 20:10, it means there are twice as many males as females, which is not true.

TIME TO TRY

Empty your pocket or purse of change. Separate the coins by denomination. Count all of the coins in each category. Now express those numbers in a ratio: ____________

Why is it important to indicate the order in which you are listing the coins? ____

____

If you did this correctly, you should have indicated the order of the coins, because without that reference, we have no idea what number corresponds with which coin. Perhaps you had 5 pennies, 4 nickels, 3 dimes, and 2 quarters. If you wrote your ratio in that order, it would be 5:4:3:2.

We also use ratios to discuss quantities in a certain amount. For example, there are about 280 million hemoglobin molecules in each red blood cell. That can be expressed as 280 million/cell, which looks more like a fraction but is read as 280 million per cell; it is really saying the ratio is 280 million to 1. Rates are also a special type of ratio. A red blood cell travels about 700 miles in its 120-day lifespan, giving a ratio of 700 miles/120 days.

Proportions are statements of equal ratios. A simple example would be to say $1/2 = 4/8$. In science, we often use proportions to solve problems. To see how, examine a generic version:

$$\frac{a}{b} = \frac{c}{d}$$

Because these two ratios are equal, their cross-products are also equal, due to some basic laws of math. This means that the product of multiplying the first numerator (a) by the second denominator (d) equals the product of multiplying the second numerator (c) by the first denominator (b):

$$a \times d = c \times b$$

Let's say we want to know how many times the heart beats in an hour. Assume that the heart beats on average 80 beats per minute. We know there are 60 minutes per hour. So, we can set up the proportion, filling in the information we know and using "x" to represent the value we are trying to determine:

$$\frac{80 \text{ beats}}{1 \text{ minute}} = \frac{x \text{ beats}}{60 \text{ minutes}}$$

We know we can **cross-multiply (Figure 1.4)**. When we do that, we get $4800 = x$, so there are 4800 beats per 60 minutes, or per hour. In fact, there is a simpler way to write this problem, which is a shorter version of cross-multiplying. We know there are 80 beats per minute, and 60 minutes per hour, so we can calculate the beats per hour as follows:

$$\frac{80 \text{ beats}}{\text{minute}} \times \frac{60 \text{ minutes}}{\text{hour}} = \frac{4800 \text{ beats } \cancel{\text{minutes}}}{\cancel{\text{minute}} \text{ hour}} = \frac{4800 \text{ beats}}{\text{hour}}$$

Notice that the units are shown, and in the next to last step, minutes appear on both the top and bottom. That means they cancel each other out, so we are left with beats per hour, which is the correct unit. Using the units can be an easy way to ensure that you have set up the problem correctly. This is another example of taking the time to think through the problem before you start—the units should make sense when you are done.

1. Set up the proportion with what is known, using "x" to represent the information you are seeking.

$$\frac{80 \text{ beats}}{1 \text{ minute}} = \frac{x \text{ beats}}{60 \text{ minutes}}$$

2. Next, cross-multiply to solve for "x."

$$\frac{80 \text{ beats}}{1 \text{ minute}} = \frac{x \text{ beats}}{60 \text{ minutes}}$$

$80 \times 60 = 4800$, and $1x = x$

So $x = 4800$ beats per 60 minutes, or 1 hour

FIGURE 1.4 Using proportions to determine how many times the heart beats in an hour.

✔ **QUICK CHECK**

An average person takes about 12 breaths per minute. How many breaths do they take in an hour? ____________

Answer: Set up the proportion: 12 breaths/minute = x breaths/60 minutes.
Cross-multiply: $12 \times 60 = 720 = x$, so $x = 720$ breaths per hour.

WHAT DID YOU LEARN?

Try these exercises from memory first, then go back and check your answers, looking up any items that you want to review. Answers to these questions are at the end of the book.

PART A: SOLVE THESE PROBLEMS.

1. $(4 \div 2) + 6 - 5 \times 2^3 =$ __________
2. $2 \times 10^4 =$ __________
3. 27/36 reduced is __________
4. $3/8 \times 2/3 =$ __________
5. $5/6 - 7/12 =$ __________
6. $0.5 \times 0.4 =$ __________
7. If the respiratory rate is 12 breaths per minute, how many breaths are taken in 1 hour? __________
8. 4 meters = __________ centimeters
9. 1 inch = 2.54 cm, so 1 foot = __________ cm
10. If a man weighs 200 pounds and 60% of his body weight is water, how many pounds of water does he have? __________

PART B: ANSWER THESE QUESTIONS.

1. What is the mean of 8, 9, 12, 18, and 23? __________
2. What is the numerator in 4/5? __________
3. Express 3/10 as a decimal __________ and as a percent __________
4. What is meant by "normal" blood pressure? __________

The BODMAS rule

When evaluating numerical expressions we need to know the order in which addition, subtraction, multiplication and division are carried out. As a simple example, consider evaluating $2 + 3 \times 4$. If the addition is carried out first we get $2 + 3 \times 4 = 5 \times 4 = 20$. If the multiplication is carried out first we get $2 + 3 \times 4 = 2 + 12 = 14$. Clearly the order of carrying out numerical operations is important. The BODMAS rule tells us the order in which we must carry out the operations of addition, subtraction, multiplication and division.

Key point

BODMAS stands for

Brackets ( )	First priority
Of ×	Second priority
Division ÷	Second priority
Multiplication ×	Second priority
Addition +	Third priority
Subtraction −	Third priority

This is the order of carrying out arithmetical operations, with bracketed expressions having highest priority and subtraction and addition having the lowest priority. Note that 'Of', 'Division' and 'Multiplication' have equal priority, as do 'Addition' and 'Subtraction'. 'Of' is used to show multiplication when dealing with fractions, for example, find $\frac{1}{2}$ of 6 means $\frac{1}{2} \times 6$.

If an expression contains only multiplication and division, we evaluate by working from left to right. Similarly, if an expression contains only addition and subtraction, we also evaluate by working from left to right.

WORKED EXAMPLES

1.1 Evaluate

(a) $2 + 3 \times 4$ (b) $(2 + 3) \times 4$

Solution

(a) Using the BODMAS rule we see that multiplication is carried out first. So

$$2 + 3 \times 4 = 2 + 12 = 14$$

(b) Using the BODMAS rule we see that the bracketed expression takes priority over all else. Hence

$$(2 + 3) \times 4 = 5 \times 4 = 20$$

1.2 Evaluate

(a) $4 - 2 \div 2$ (b) $1 - 3 + 2 \times 2$

Solution (a) Division is carried out before subtraction, and so

$$4 - 2 \div 2 = 4 - \frac{2}{2} = 3$$

(b) Multiplication is carried out before subtraction or addition.

$$1 - 3 + 2 \times 2 = 1 - 3 + 4 = 2$$

1.3 Evaluate

(a) $(12 \div 4) \times 3$ (b) $12 \div (4 \times 3)$

Solution Recall that bracketed expressions are evaluated first.

(a) $(12 \div 4) \times 3 = \left(\frac{12}{4}\right) \times 3 = 3 \times 3 = 9$

(b) $12 \div (4 \times 3) = 12 \div 12 = 1$

Example 1.6 shows the importance of the position of brackets in an expression.

Self-assessment questions 1.1

1. State the BODMAS rule used to evaluate expressions.
2. The position of brackets in an expression is unimportant. True or false?

Exercise 1.1

1. Evaluate the following expressions:
   (a) $6 - 2 \times 2$ (b) $(6 - 2) \times 2$
   (c) $6 \div 2 - 2$ (d) $(6 \div 2) - 2$
   (e) $6 - 2 + 3 \times 2$ (f) $6 - (2 + 3) \times 2$
   (g) $(6 - 2) + 3 \times 2$ (h) $\frac{16}{-2}$ (i) $\frac{-24}{-3}$
   (j) $(-6) \times (-2)$ (k) $(-2)(-3)(-4)$

2. Place brackets in the following expressions to make them correct:
   (a) $6 \times 12 - 3 + 1 = 55$
   (b) $6 \times 12 - 3 + 1 = 68$
   (c) $6 \times 12 - 3 + 1 = 60$
   (d) $5 \times 4 - 3 + 2 = 7$
   (e) $5 \times 4 - 3 + 2 = 15$
   (f) $5 \times 4 - 3 + 2 = -5$

Prime numbers and factorisation

A **prime number** is a positive integer, larger than 1, which cannot be expressed as the product of two smaller positive integers. To put it another way, a prime number is one that can be divided exactly only by 1 and itself. For example, $6 = 2 \times 3$, so 6 can be expressed as a product of smaller numbers and hence 6 is not a prime number. However, 7 is prime. Examples of prime numbers are 2, 3, 5, 7, 11, 13, 17, 19, 23. Note that 2 is the only even prime.

Factorise means 'write as a product'. By writing 12 as 3×4 we have factorised 12. We say 3 is a **factor** of 12 and 4 is also a factor of 12. The way in which a number is factorised is not unique: for example, 12 may be expressed as 3×4 or 2×6. Note that 2 and 6 are also factors of 12.

When a number is written as a product of prime numbers we say the number has been **prime factorised**.

To prime factorise a number, consider the technique used in the following examples.

WORKED EXAMPLES

1.4 Prime factorise the following numbers:

(a) 12 (b) 42 (c) 40 (d) 70

Solution

(a) We begin with 2 and see whether this is a factor of 12. Clearly it is, so we write

$$12 = 2 \times 6$$

Now we consider 6. Again 2 is a factor so we write

$$12 = 2 \times 2 \times 3$$

All the factors are now prime, that is the prime factorisation of 12 is $2 \times 2 \times 3$.

(b) We begin with 2 and see whether this is a factor of 42. Clearly it is and so we can write

$$42 = 2 \times 21$$

Now we consider 21. Now 2 is not a factor of 21, so we examine the next prime, 3. Clearly 3 is a factor of 21 and so we can write

$$42 = 2 \times 3 \times 7$$

All the factors are now prime, and so the prime factorisation of 42 is $2 \times 3 \times 7$.

(c) Clearly 2 is a factor of 40,

$$40 = 2 \times 20$$

Clearly 2 is a factor of 20,

$$40 = 2 \times 2 \times 10$$

Again 2 is a factor of 10,

$$40 = 2 \times 2 \times 2 \times 5$$

All the factors are now prime. The prime factorisation of 40 is $2 \times 2 \times 2 \times 5$.

(d) Clearly 2 is a factor of 70,

$$70 = 2 \times 35$$

We consider 35: 2 is not a factor, 3 is not a factor, but 5 is.

$$70 = 2 \times 5 \times 7$$

All the factors are prime. The prime factorisation of 70 is $2 \times 5 \times 7$.

1.5 Prime factorise 2299.

Solution We note that 2 is not a factor and so we try 3. Again 3 is not a factor and so we try 5. This process continues until we find the first prime factor. It is 11:

$$2299 = 11 \times 209$$

We now consider 209. The first prime factor is 11:

$$2299 = 11 \times 11 \times 19$$

All the factors are prime. The prime factorisation of 2299 is $11 \times 11 \times 19$.

Self-assessment questions 1.2

1. Explain what is meant by a prime number.
2. List the first ten prime numbers.
3. Explain why all even numbers other than 2 cannot be prime.

Exercise 1.2

1. State which of the following numbers are prime numbers:
(a) 13 (b) 1000 (c) 2 (c) 29 (d) $\frac{1}{2}$
2. Prime factorise the following numbers:
(a) 26 (b) 100 (c) 27 (d) 71 (e) 64 (f) 87 (g) 437 (h) 899
3. Prime factorise the two numbers 30 and 42. List any prime factors which are common to both numbers.

Chapter 2
Measurement

03

Measurement

Objectives: This chapter:

- revises common units of length and conversions between them
- revises the concepts of area and volume
- provides a summary of important formulae needed for calculating areas of common shapes and volumes of common solids
- revises common units of mass and conversions between them

Units of length, area and volume

The standard unit of **length** is the metre, abbreviated to m, although several smaller units are commonly used as well. The following relationships are given for future reference:

Key point

1 metre (m) = 100 centimetres (cm)
= 1000 millimetres (mm)

Note also that 1000 metres is equal to 1 kilometre (km).

WORKED EXAMPLES

2.1 Convert a length of 6.7 metres into centimetres.

Solution Because 1 m equals 100 cm then 6.7 metres equals $6.7 \times 100 = 670$ cm.

2.2 Convert a length of 24 mm into metres.

Solution Because 1 m equals 1000 mm, then 1 mm equals $\frac{1}{1000}$ m. Therefore 24 mm equals

$$24 \times \frac{1}{1000} = 0.024 \text{ m}$$

The **area** of a shape is measured by finding how many square units it contains. 1 square metre, written 1 m^2, is the area of a square having sides of length 1 metre. Similarly 1 square centimetre, written 1 cm^2, is the area of a square having sides of length 1 centimetre.

The **volume** of a solid is measured by finding how many cubic units it contains. 1 cubic metre, written 1 m^3, is the volume of a cube with sides of length 1 metre. Similarly, 1 cubic centimetre, written 1 cm^3, is the volume of a cube having sides of length 1 centimetre.

When measuring the volume of liquids, the litre is commonly used. One litre, that is, 1 l, is equal to 1000 cubic centimetres, that is,

$$1 \text{ l} = 1000 \text{ cm}^3$$

Note that 1000 cm^3 is the volume of a cube of side 10 cm, and so it is helpful to remember that such a cube holds 1 litre of liquid. Commonly millilitres are used. 1 millilitre (1 ml) equals 1 cubic centimetre, that is,

$$1 \text{ ml} = 1 \text{ cm}^3$$

Key point

1 litre (l) = 1000 millilitres (ml)

1000 litres (l) = 1 m^3

WORKED EXAMPLE

2.3 Convert 256 millilitres into litres.

Solution Because 1 l = 1000 ml then 1 ml is equal to $\frac{1}{1000}$ l. Therefore

$$256 \text{ ml} = 256 \times \frac{1}{1000} = 0.256 \text{ l}$$

The areas of common shapes and volumes of common solids can be found using formulae. Table 2.1 summarises a variety of such formulae. The calculation of an area or volume is then simply an exercise in substituting values into a formula.

Table 2.1
Formulae for areas and volumes of common shapes and solids

Rectangle	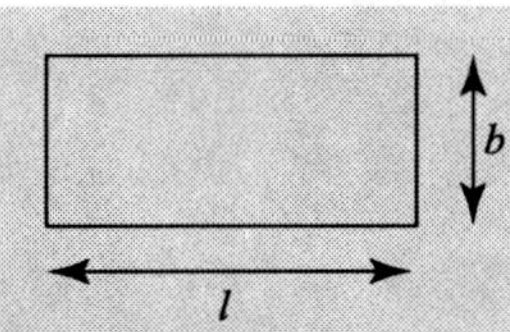	Area $= l \times b$
Parallelogram	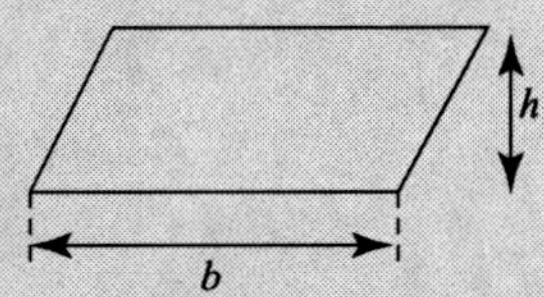	Area $= b \times h$
Triangle	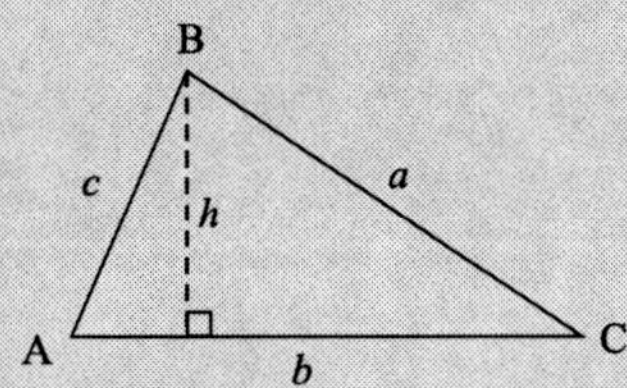	Area $= \frac{1}{2}bh$ Area $= \sqrt{s(s-a)(s-b)(s-c)}$ where $s = \dfrac{a+b+c}{2}$
Trapezium	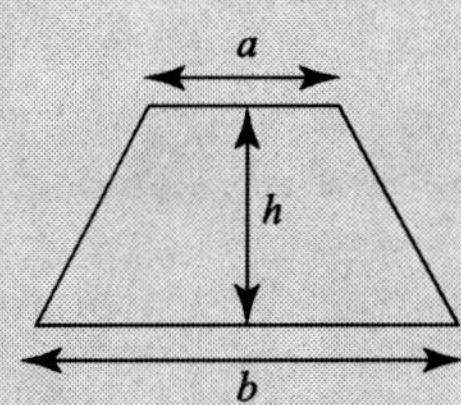	Area $= \frac{1}{2}(a+b)h$
Circle	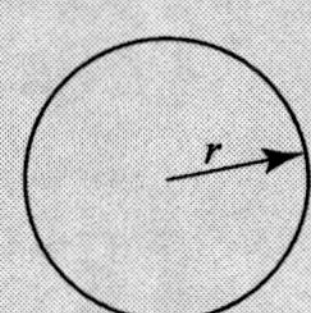	Area $= \pi r^2$ Circumference $= 2\pi r$ $(\pi = 3.142)$
Sector of a circle		Area $= \frac{1}{2}r^2\alpha$ (α in radians)
Rectangular box	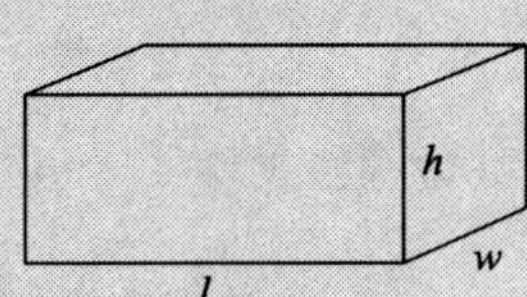	Volume $= l \times h \times w$
Cylinder	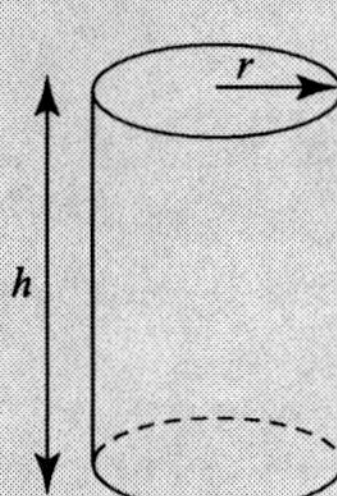	Volume $= \pi r^2 h$ Curved surface area $= 2\pi rh$

(continued)

Table 2.1
Continued

Cone	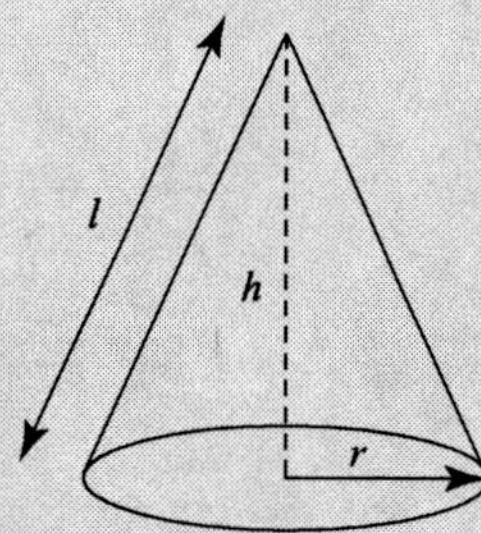	Volume $= \frac{1}{3}\pi r^2 h$ Curved surface area $= \pi r l$
Sphere	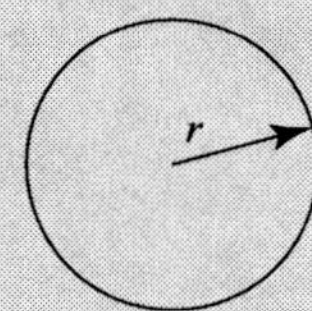	Volume $= \frac{4}{3}\pi r^3$ Surface area $= 4\pi r^2$

WORKED EXAMPLES

2.4 A square has sides of length 1 m. Calculate the area in

(a) m^2 (b) cm^2 (c) mm^2

Solution

(a) When working in metres the area is $1 \times 1 = 1\ m^2$.

(b) When working in centimetres, each side is of length 100 cm. Therefore the area is $100 \times 100 = 10\,000 = 10^4\ cm^2$.

(c) When working in millimetres, each side is of length 1000 mm and so the area is $1000 \times 1000 = 10^3 \times 10^3 = 10^6\ mm^2$.

Note that in all three cases the areas are the same and so we can say

$$1\ m^2 = 10^4\ cm^2 = 10^6\ mm^2$$

2.5 A cubic box has sides of length 1 m. Calculate its volume in

(a) m^3 (b) cm^3 (c) mm^3

Solution

(a) When working in metres the volume is $1 \times 1 \times 1 = 1\ m^3$.

(b) When working in centimetres, each side is of length 100 cm. Therefore the volume is $100 \times 100 \times 100 = 1\ 000\ 000 = 10^6\ cm^3$.

(c) When working in millimetres, each side is of length 1000 mm and so the volume is $1000 \times 1000 \times 1000 = 10^3 \times 10^3 \times 10^3 = 10^9\ mm^3$.

Note that in all three cases the volumes are the same and so we can say

$$1\ m^3 = 10^6\ cm^3 = 10^9\ mm^3$$

Exercise 2.1

1. Convert 16 cm into (a) mm (b) m.
2. Change 356 mm into (a) cm (b) m.
3. Change 156 m into km.
4. Find the area of a rectangle whose length is 14 cm and whose width is 7 cm.
5. A room has length 5 m and width 3.5 m. It is required to carpet this room at £4.99 per square metre. Calculate the cost of the carpet assuming there is no wastage due to cutting.
6. Find the area of a parallelogram that has a vertical height of 2 m and a base of length 8 m.
7. A triangle has sides of length 8 cm, 12 cm and 14 cm. Find its area.
8. Find the area of a trapezium whose parallel sides are 14 cm and 8 cm long and whose distance apart is 2 cm.
9. Find the area of a circle whose radius is 18 cm.
10. Find the area of a circle whose diameter is 10 cm.
11. Find the radius of the circle whose area is 26 cm^2.
12. A circular piece of cardboard of radius 70 cm has a circular piece of radius 25 cm removed from it. Calculate the area of the remaining cardboard.
13. Find the volume of a tin of beans of base radius 5 cm and height 14 cm. What is the tin's capacity in litres?
14. Find the volume of a sphere of radius 8 cm.
15. A sphere has volume 86 cm^3. Calculate its radius.
16. Calculate the area of a sector of a circle of angle 45° and radius 10 cm.
17. A garden swimming pool is designed in the form of a cylinder of base radius 3 m. Calculate how many litres of water are required to fill it to a depth of 1 m.

Units of mass

The standard unit of mass is the kilogram, abbreviated to kg. Sometimes smaller units, the gram (g) or milligram (mg), are used. A very sensitive balance may be able to measure in micrograms (μg). The relationship between these units is given as follows:

Key point

1 kilogram (kg) = 1000 grams (g) = 1 000 000 milligrams (mg)
1 gram (g) = 1000 milligrams (mg) = 1 000 000 micrograms (μg)
1 milligram = 1000 micrograms (μg)

Note also that 1000 kilograms equals 1 tonne (t).

WORKED EXAMPLES

2.6 Convert 8.5 kg into grams.

Solution 1 kg is equal to 1000 g and so 8.5 kg must equal $8.5 \times 1000 = 8500$ g.

2.7 Convert 50 g to kilograms.

Solution Because 1 kg equals 1000 g then 1 g equals $\frac{1}{1000}$ of a kilogram. Consequently,

$$50 \text{ g} = 50 \times \frac{1}{1000} = 0.05 \text{ kg}$$

Exercise 2.2

1. Convert 12 kg into (a) grams (b) milligrams.
2. Convert 168 mg into (a) grams (b) kilograms.
3. Convert 0.005 t into kilograms.
4. Convert 875 t into kilograms.
5. Convert 3500 kg into tonnes.

Test and assignment exercises 2

1. Change the following to grams:
   (a) 125 mg (b) 15 kg (c) 0.5 mg
2. Change the following lengths to centimetres:
   (a) 65 mm (b) 0.245 mm (c) 5 m
3. Convert 7500 m and 125.5 m into kilometres.
4. Convert 8 l into millilitres.
5. Convert 56 ml into litres.
6. A household water tank has dimensions $1 \text{ m} \times 0.6 \text{ m} \times 0.8$ m. Calculate its capacity in litres assuming it can be filled to the brim.
7. Calculate the area of the sector of a circle of radius 25 cm and angle 60°.
8. A sector of a circle of radius 10 cm has area 186 cm^2. Calculate the angle of the sector.
9. Calculate the volume of a cylinder whose height is 1 m and whose base radius is 10 cm. (Be careful of the units used.)
10. A sphere has diameter 16 cm. Calculate its volume and surface area.

11. A trapezium has parallel sides of length y_1 and y_2. If these sides are a distance h apart find an expression for the area of the trapezium. Calculate this area when $y_1 = 7$ cm, $y_2 = 10$ cm and $h = 1$ cm.

12. A lorry weighs 7.5 t. Express this in kilograms.

Measurement in biology

Units

Measurements, when dealing with dimensions, don't mean anything without units. Units are vital in communicating your results and helping you work out if you are calculating something properly.

You need to know these prefixes:

giga-	10^9
mega-	10^6
kilo-	10^3
1	1
milli-	10^{-3}
micro-	10^{-6}
nano-	10^{-9}
pico-	10^{-12}
femto	10^{-15}

SI units

There are seven units, known as SI units, from which all others are derived:

Length	meter	m
Time	second	s
Mass	kilogramme	kg
Electric current	ampere	A
Thermodynamic temperature	kelvin	K
Amount of substance	mole	mol
Luminous intensity	candela	cd

These are the universally accepted units of science.
Other concepts have units with their base in the SI units. Examples are:

Area	m^2
Volume	m^3
Density	$kg\ m^{-3}$
Velocity, speed	$m\ s^{-1}$
Acceleration	$m\ s^{-2}$

Ideally you would work in SI units, and always express numbers in their SI units. However, common usage means that, for example, grams are often used instead of kg, and minutes or hours instead of seconds; this is acceptable. However, it is preferable to avoid quoting answers in units such as cm when presenting data in a results section, on graphs or tables for example.

How to avoid lots of zeros when converting to SI units

If you carry out a calculation and the answer is, for example 600 km and you wanted to convert this to the SI units m:

1000 m = 1 km

So you would multiply 600 m by 1000 giving 600 000 m.

Rather that use all these zeros, you can express 600 000 as 6 x 10^5 m.

For further explanation, see the section on scientific notation.

Problems may ask you to give the answer in units others than those in the question. To avoid making mistakes, only convert between units at the beginning or end of a calculation, NEVER in the middle

An example: An elephant walked 30 m in 12 h, how many km will it walk in 20 days?

First of all the distance:
The question is in m, the answer is asked for in km. So you need to convert 30m into km.

There are 1000 m in 1 km, or 0.001 km = 1m.

So 30 m needs to be multiplied by 0.001 to convert it to km, 30m = 0.03 km

Secondly the time:
You need to convert 12 h into days.

There are 24 hours in one day, so $\frac{12}{24} = 0.5\ days$

Then you need to look at the question. It is like a proportion. If the elephant walks 0.03 km in 0.5 days, how many multiples of 0.03km will it walk in 20 days.

So how many 0.5 days fit in 20 days, or $\frac{20\ days}{0.5\ days}$

It is this multiple of 0.03 km, so: $\frac{20\ days}{0.5\ days} \times 0.03\ km = 1.2\ km$

The answer is: the elephant walks 1.2 km in 20 days.

You need to be able to convert between units with ease, and should practise if you do not find it easy. You can find problems at the end of this chapter and also on Blackboard.

Working in areas

You may need to calculate the area of a leaf with which you are measuring photosynthesis, or the area of tissue you have stained, or the area over which you are counting bacteria colonies. Most biological shapes are irregular, but it is acceptable to make an approximation to standard shapes. For example:

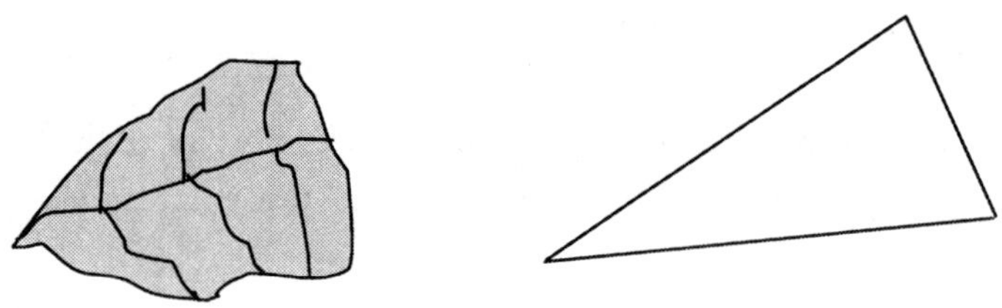

Figure 2.1
Approximating the area of a leaf to a triangle

this leaf can approximate to a triangle.

See table of areas and volumes for a list of shapes and the formulae for their areas.

For most though, the problem comes when converting between units in areas. It is not difficult. Most make mistakes by trying to make conversions without properly working it out. Avoid short cuts, and you won't go wrong If you are faced with any conversion that involves an area, draw yourself a square and work out the conversion, don't try and guess.

For example, convert 0.004 km^2 to m^2

First work out how many m^2 in 1 km^2:

.

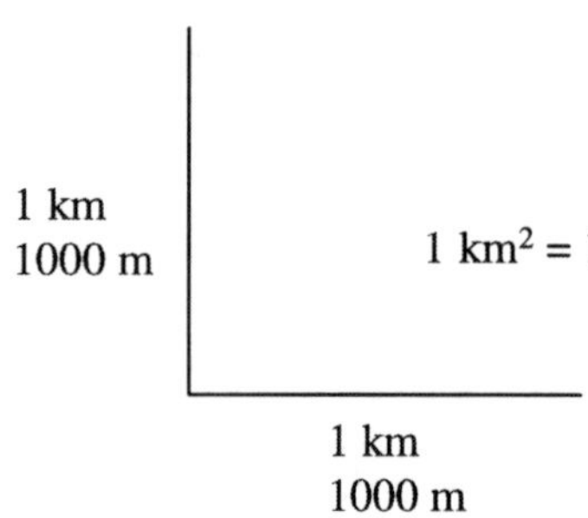

Figure 2.2
Conversion between m and km

$1\ km^2 = 1000\ m \times 1000\ m = 1000000\ m^2$

So you need to multiply 0.004 km^2 by 1000 000 = 4000 m^2 or $4 \times 10^3\ m^2$

Example: If you found 2 beetles in an area 4 x 10^4 mm^2, how many would you expect to find in 1 m^2?

Firstly, 4 x 10^4 mm^2 is 40 000 mm^2, or 200 mm x 200mm.
The question is asking how many of these 200 x 200 mm^2 would you get in 1 m^2. When in doubt, draw yourself a diagram.

Figure 2.3 Converting between areas of different sizes

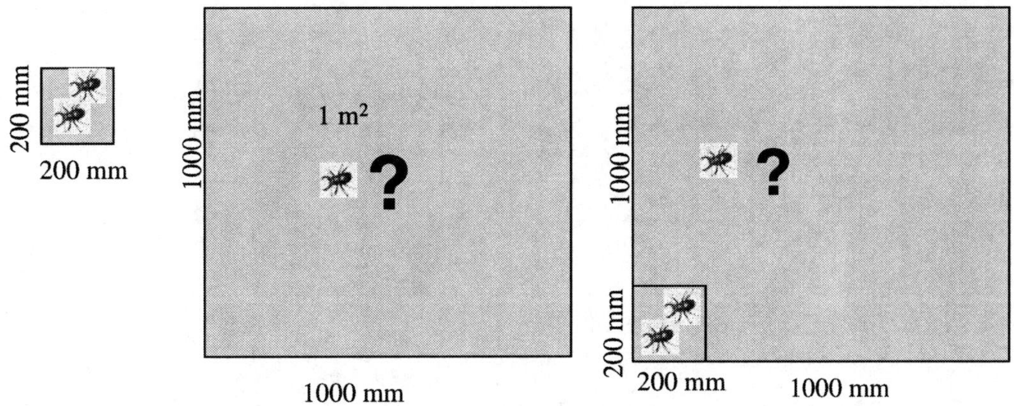

1 m^2 = 1000 000 mm^2 or 1 x 10^6 mm

Figure 2.4 Relating a small area to a larger area

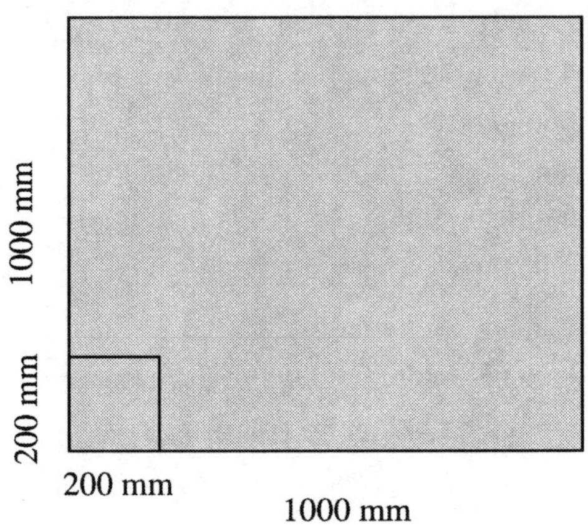

The big area is $\frac{1000\ mm}{200\ mm}$ x $\frac{1000\ mm}{200\ mm}$ bigger than the smaller area.

$$= \frac{1000000\ mm^2}{40000\ mm^2} = 25\ times\ bigger$$

If the area is 25 times bigger, there must be 25x more beetles in the larger area compared to the smaller area, so multiply the number of beetles by 25:

2 x 25 = 50 beetles.

Answer: You would expect to find 50 beetles in 1 m^2.

You can rewrite this as the equation:

$$\frac{large\ area}{small\ area} \times number\ in\ small\ area = number\ in\ large\ area$$

Plugging in the numbers from above (being careful to get the units right):

$$\frac{1000000\ mm^2}{40000\ mm^2} \times 2\ beetles = 50\ beetles$$

Calculating areas

The area of a square, for example the area of a quadrat:

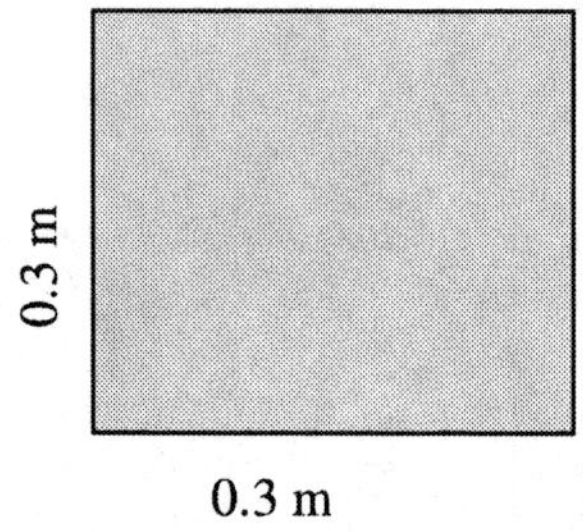

Figure 2.5
Square

$$Area = 0.3\,m \times 0.3\,m = 0.9\,m^2$$

Area of a triangle, for example, the area of a leaf:

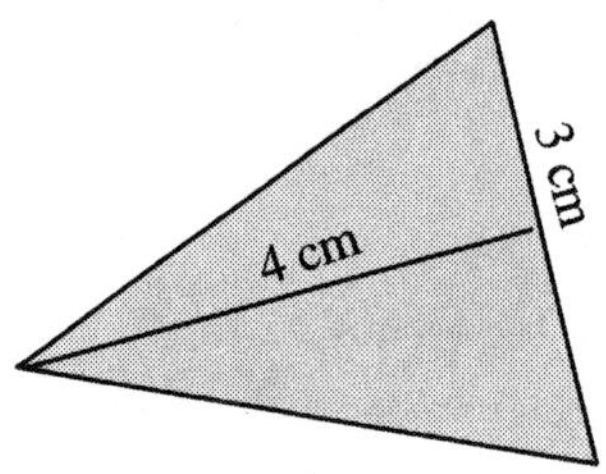

Figure 2.6
Triangle

$$Area = 0.5\,(base \times height)$$

$$Area = 0.5\,(3 \times 4) = 6\ cm^2$$

Area of a circle, for example, the field of view under a microscope:

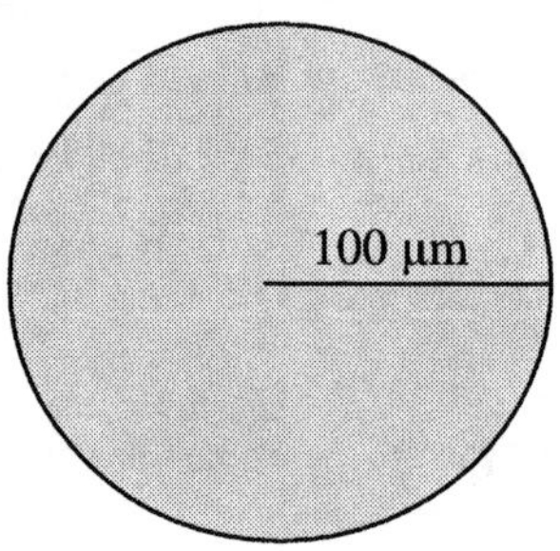

Figure 2.7
Circle

$$Area = \pi\,r^2$$

$$Area = 3.142 \times 100\mu m \times 100\mu m$$

$$Area = 31420\mu m^2$$

Note that with circles it is rarely the radius that is measured. This is because it is almost impossible to estimate the centre of a circle accurately. You will usually be measuring the circumference or the diameter of the circle.
If the diameter is measured, the radius is simply half the diameter.
If the circumference is measured the radius can be determined by rearranging the equation $C = 2\pi r$ where C is the circumference:

$$r = \frac{C}{2\pi}$$

Working in volumes

You may need to calculate the volume of a cell so that you can determine whether cell volume changes with a change in experimental conditions, or you may need to calculate the volume of a tree trunk to determine increase in biomass over time.

As with areas, you can make approximations to geometric shapes. For example many protist cells approximate to a sphere and many plant cells approximate to a cuboid.

Three dimensional shapes and their formulae can be found on page 35

As with areas, mistakes are most commonly made when converting between units.

For example: If the volume of a plant cell is 8000 μm^3, how many cells would be present in a volume of leaf $1mm^3$?

First convert the units, converting mm^3 to μm^3. Don't try and guess; draw yourself a diagram:

Figure 2.8
Converting units for volumes

$1mm^3$=1000mm x 1000 mm x 1000 mm

$1mm^3 = 1000\,000\,000\ \mu m^3$

Or $1 \times 10^9\ \mu m^3$

The problem is one of proportions, being the proportion of the big volume over the smaller volume:

$$\frac{Big\ volume}{Little\ volume} \times number\ of\ \ plant\ cells\ in\ little\ volume = number\ of\ \ plant\ cells\ in\ big\ volume$$

$$\frac{1000000000\,\mu m^3}{8000\,\mu m^3} \times 1\,plant\ cell = 125000\,plant\ cells$$

With all problems such as this, it is really important to look at your answer and look back at the question and check the answer makes sense.

Volume conversions can apply to liquids too. For example: A biologist has put 0.1 g salt into 125 ml water. What is the concentration of salt in gl^{-1}?

The small volume is 125 ml, the big volume is 1 litre. The first thing is to convert units, converting ml into litres. Don't guess – draw a diagram:

1 ml = 1 cm^3

1 litre is a volume 10 cm x 10 cm x 10 cm = 1000 cm^3

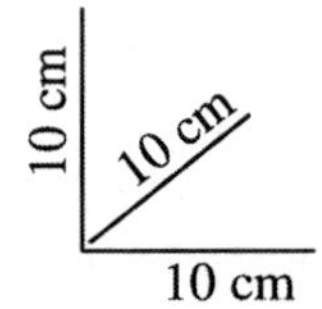

So 1 litre = 1000 ml

$$\frac{Big\ volume}{Little\ volume} \times mass\ in\ little\ volume = mass\ in\ big\ volume$$

$$\frac{1000ml}{125ml} \times 0.1g = 0.8g$$

The answer is: the concentration of salt would be 0.8 g in 1 litre, or 0.8 g l^{-1}.

Calculating volumes

A plant cell:

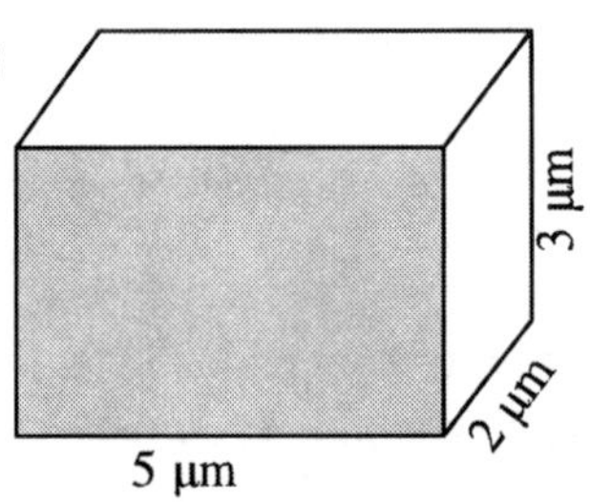

Figure 2.9 Diagrammatic representation of a plant cell

$$Volume\ of\ the\ cell = 5 \times 2 \times 3 = 30\mu m^3$$

A protist cell:

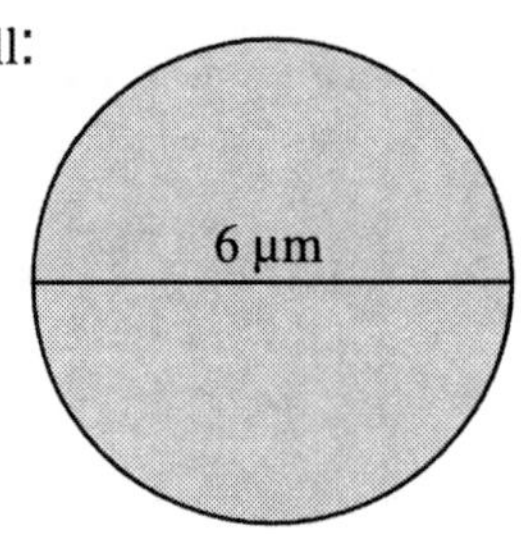

Figure 2.10 An approximate shape for a spherical protist cell

$$Volume\ of\ a\ sphere = \frac{4}{3}\pi r^3$$

The radius is half the diameter:

$$\frac{6\mu m}{2} = 3\mu m$$

$$Volume = \frac{4 \times 3.142 \times 3\mu m \times 3\mu m \times 3\mu m}{3}$$

$$Volume\ of\ the\ cell = 113.04\mu m^3$$

Be careful with these calculations because you are dealing with areas and volumes and when converting between units remember to convert the units properly; avoid short cuts.

Test your understanding.

Carry out the following conversions:

1. Convert to mm:

a. 1200 µm

b. 0.4 cm

c. 0.0035 m

2. Convert to g:

a. 0.045 kg

b. 36 mg

c. 64000µg

3. Convert to m^2:

a. 3500 mm^2

b. 480000 μm^2

c. 0.0005 km^2

4. Convert to mm^3

a. 0.000004 m^3

b. 853000 μm^3

c. 0.004 cm^3

5. Convert to ml:

a. 34 cm^3

b. 5 l

c. 680 µl

6. Calculate the following:

a. If a squirrel gathered 18 acorns in 30 min, now many could it gather in 3 hours?

b. If a camel walked 11.6 km across a desert in one day, how far would it walk in 25 days?

c. If you had 0.5 g of KCl in 40 ml, what would be the concentration of KCl in g l^{-1}?

d. There is 1 acacia tree in every 600 m^2 of grassland. How many might there be in 4 km^2?

e. If there were 3 cod fish in 1000 litres of sea how many would there be in 4 km^3?

f. If there were 4×10^3 woodlice in 10m^2 of leaf litter how many would there be in 1 km^2?

7. Calculate the following:

a. A spherical cell has a diameter of 12 µm, what is the volume of this cell?

b. A caterpillar has a circumference of 16 mm and a length of 24 mm. What is the volume of the caterpillar?

c. A leaf, approximating to a triangle, has a length of 38 mm and a base width of 29 mm. What is its area?

Calculate:

8. The area of a square with sides of 4.2 mm.

9. The area of a quadrat with sides of 0.56 m.

10. The volume of a cube with sides of 4.3 cm.

11. The area of a circle with a radius of 7 mm.

12. The area of a circle with a diameter of 2.9 m.

13. The area of a circle with a circumference of 1.6 μm.

14. The volume of a sphere with a radius of 5.75 m.

15. The volume of a sphere with a circumference of 0.23 m.

16. The volume of a sphere with a diameter of 5 km.

17. The volume of a cylinder with a radius of 7 μm and a length of 10 μm.

18. The volume of a cylinder with a circumference of 3.7 m and a height of 6.6 m.

19. The volume of a cone with a height of 4.1 cm and a radius of 2.9 cm.

20. The biomass of a boletus mushroom, which has a top part that approximates to a half sphere. The mushroom tissues weigh 0.03 g cm^{-3}. The mushroom stalk is 4.0 cm high, and has a circumference of 4.5 cm. The top of the mushroom has a diameter of 6.8 cm.

21. The mass of a spherical protist cell which has a diameter of 32 μm . The cell has a mass per unit volume of 3 ng μm^{-3}. Calculate the mass of the cell.

22. The volume of a bacterial colony that contains 5×10^8 bacteria, where each bacterial cell approximates to a cylinder and has a length of 0.84 μm and a diameter of 0.33 μm.

23. The mass of a flatworm that has a length of 0.89 m, a thickness of 1mm and a width of 5 mm. It has a mass per unit volume of 0.4 g cm^{-3}.

24. The total volume of earthworms in a field 0.4 km^2. Each worm approximates to a cylinder 7 cm long and 8mm in circumference and there are 8 worms per m^2.

Chapter 3
Moles and Molarity

Moles and Molarity

What is a mole?

A mole of anything, could be glucose, or salt, or tennis balls, contains 6.022×10^{23} individuals. For glucose it would be 6.022×10^{23} glucose molecules, for salt it would be 6.022×10^{23} salt molecules, for tennis balls it would be 6.022×10^{23} tennis balls. In all these instances, each individual has a mass, in grams. The mass of a salt molecule is extremely tiny, but the mass of 1 mole of salt molecules is 58.44 g. This is a useful amount to work with. A mole of glucose molecules would weigh 180.18 g. A mole of tennis balls would be extremely heavy:

One tennis ball has a mass of 55.15 g

6.022×10^{23} tennis balls have a mass of 3.31×10^{25} g (approximately half the mass of the moon)

The number of individuals in 1 mole is called Avogadro's number = 6.022×10^{23}

What is molar mass?

The mass of one mole of anything is referred to as the molar mass and has units of grams per mole. It is the mass of 6.022×10^{23} individuals. The individuals referred to are molecules in the examples below.
The molar mass of NaCl (6.022×10^{23} molecules of NaCl) is 58.44g mol^{-1}. So to get 1 mol of NaCl you would have to weigh out 58.44g of NaCl.
The molar mass of ZnCl (6.022×10^{23} molecules of ZnCl) is 136 g mol^{-1}. So to get 1 mol of ZnCl you would have to weigh out 136g of ZnCl.

How do you work out the molar mass?

First you have to know the mass of one molecule of the substance you are interested in. To do this you need to know the mass of the atoms making up the molecule. You add together the masses of all the protons and neutrons in the atom (electrons have a negligible mass) and this gives you the mass of the atom. This is going to be an extremely small number. You then add up the masses of the atoms in one molecule. Then this can be multiplied by Avogadro's number to give you the mass of one mole of that molecule. For example:

Take the carbon atom ^{12}C. It has 6 protons and 6 neutrons.

The mass of a proton is 1.6726×10^{-24} g

The mass of a neutron is 1.6750×10^{-24} g

So, to take 6 protons and 6 neutrons, you multiply each of the above numbers by 6.

That is: $(1.6726 \times 10^{-24}\ g \times 6) + (1.6750 \times 10^{-24}\ g \times 6)$

you get $10.0356 \times 10^{-24}\ g + 10.0500 \times 10^{-24}\ g = 20.0856 \times 10^{-24}\ g$

Multiply this by Avogadro's number:

20.0856×10^{-24} g x 6.022×10^{23} = 12.0955 g

So the mass in g of a mole of ^{12}C atoms is 12.0955 g.

Or, the molar mass of ^{12}C is 12.0955 g

What Avogadro's number does

Avogadro's number converts the very tiny mass of atoms and molecules into a usable number. The number each molecule is converted to, its molar mass, is also approximately equal to the number of protons and neutrons in the molecule. For example, Avogadro's number converts the mass of a mole of ^{12}C to approximately 12 g, and this is very useful because there are 6 neutrons and 6 protons in ^{12}C. This has then been referred to as 12 atomic mass units (amu), equivalent to the number of neutrons and protons in the atom ^{12}C.

The conversion can be applied to the atoms themselves. All atoms and molecules are given a mass relative to one twelfth the mass of ^{12}C, taking the mass of ^{12}C to be 12 amu. This unit is now called the Dalton (Da). By using this conversion, protons and neutrons both have a mass of approximately 1 amu, or 1 Da. This is very convenient because you can work out the number of Da for any molecule, and this amount is equivalent to the number of protons and neutrons in the molecule, and also its molar mass in grams.

Terms you will come across, and what they mean

Molecular mass or molecular weight

Text books use these terms a lot. It means the number of Daltons that make up a molecule. It does not refer to their actual mass in grams, it refers to their mass relative to ^{12}C, as described above.

Relative molecular mass (M_r)

This is similar to the molecular mass, in that it is the mass of a molecule relative to the mass of one twelfth the mass of ^{12}C. Since ^{12}C is taken to be 12 Da, one twelfth of 12 Da is 1 Da. Because molecular mass is in Daltons, and one twelfth the mass of ^{12}C is 1 Da, the units cancel out. So M_r is the same number as molecular mass, but there are no units. Relative molecular mass is what you will find on any container of a chemical (although some list this as Mwt, or molecular weight). The number is equivalent to the molar mass (but molar mass has units of g mol^{-1}), so is a very useful number to know for any chemical.

Dalton

The Dalton is a derived unit, very useful to molecular biologists. You can work out the number of Daltons that make up a molecule, just by adding up the number of atoms in the molecule, or taking the molar mass and replacing g mol^{-1} with Da.

All these terms and their units are derived. The only expression that is not derived is molar mass. Molar mass is a conversion of atomic mass, using Avogadro's number and has the units of grams per mole. For example:

NaCl — Molar mass = 58.44 g mol^{-1}

Na atomic mass = 22.99 Da

Cl atomic mass = 35.45 Da

Molecular mass = 58.44 Da

Relative molecular mass = 58.44

Glucose $C_6H_{12}O_6$ — Molar mass = 180.18 g mol^{-1}

C atomic mass = 12.01 Da

H atomic mass = 1.01 Da

O atomic mass = 16.00 Da

Molecular mass = (12.01 x 6)+(1.01 x 12)+(16.00 x 6) = 180.18 Da

Relative molecular mass = 180.18

Why work in moles?

Knowledge of the number of moles in a compound gives a direct indication of the number of molecules in that compound. For example, if you take 6.022×10^{23} molecules of NaOH and put them with 6.02×10^{23} molecules of HCl, all the molecules will react together. Where as if you took 1 g of each, there would be molecules of HCl left over because in 1 g NaOH there would be 1.51×10^{22} molecules, and in 1 g of HCl there would be 1.65×10^{22} molecules. So it is the number of moles that is important when looking at chemicals reacting together, because that way you look at the number of molecules of each chemical.

Converting between mass and moles

The mass of a substance (g) is equal to the sum of the molar mass (g mol^{-1}) multiplied by the number of moles (mol).

$$mass(g) = molar\ mass(g\ mol^{-1}) \times moles(mol)$$

To calculate the number of moles in a sample, the equation can be rearranged to give:

$$moles(mol) = \frac{mass(g)}{molar\ mass(g\ mol^{-1})}$$

For example, you may see the question: How many moles of NaCl are there in 5.844g of NaCl? Molar mass of NaCl = 58.44 g mol^{-1}.

Here you would take the equation in the arrangement where you want to know the number of moles (mol):

$$moles(mol) = \frac{mass(g)}{molar\ mass(g\ mol^{-1})}$$

Answer: there are 0.1 moles of NaCl in 5.844 g of NaCl.

$$moles(mol) = \frac{5.844\ g}{58.44\ g\ mol^{-1}} = 0.1\ mol$$

Molarity

Molarity is a term used when dealing with solutions. The molarity of a solution refers to the number of moles in one litre, so it is referring to the concentration of a solution. It has the units of mol l^{-1}.

$$molarity\ (mol\ l^{-1}) = \frac{number\ of\ moles\ (mol)}{volume\ (l)}$$

So you could be asked, what is the molarity of a solution with 3 moles of NaCl in 0.5l?

Taking the above equation, you would substitute the numbers:

$$molarity\ (mol\ l^{-1}) = \frac{3\ mol}{0.5\ l} = 6\ mol\ l^{-1}$$

Answer: the molarity of a 0.5 l solution with 3 moles of NaCl is 6 mol l^{-1}

Note – you do not need the molar mass of NaCl in this equation.

If you were asked the molarity of a 2l solution with 3 g of NaCl in it, you can combine two equations:

$$moles(mol) = \frac{mass(g)}{molar\ mass(g\ mol^{-1})} \qquad molarity\ (mol^{-1}) = \frac{number\ of\ moles\ (mol)}{volume\ (l)}$$

Both equations have moles in them, so the first equation can substitute the moles term in the second equation. By subsitiuting the first equation into the number of moles in the second equation you get:

$$molarity\ (mol\ l^{-1}) = \frac{mass(g)}{molar\ mass\ (g\ mol^{-1}) \times volume\ (l)}$$

Note that in this case you do need to know the molar mass of NaCl, which is 58.44 g mol^{-1}

$$molarity\ \left(mol\ l^{-1}\right) = \frac{3g}{58.44\ g\ mol^{-1} \times 2\ l} = 0.026\ mol\ l^{-1}$$

Answer: the molarity of a 2l solution containing 3 g NaCl is 0.026 mol l^{-1}
Note that units of molarity are sometimes expressed as M, and you will see this in many text books. In this book we use mol l^{-1} so that units can always be solved in equations.

Making and diluting solutions

Figure 3.1
Dilution of a substance

If you take a chemical and dilute it, the amount (moles) of chemical before and after dilution is the same, but the concentration of the chemical has changed. Look at the following equation:

$$moles\ (mol) = molarity\ (mol\ l^{-1}) \times volume\ (l)$$

If the moles stay the same, but the volume has increased, to keep the sides of the equation equal to each other, the molarity must decrease.
If you put a teaspoon of sugar in a cup of tea, and then add more tea, the tea still contains one teaspoon of sugar, but it will taste less sweet. This is because the concentration, or molarity, of the sugar has gone down as the volume goes up.

If you apply this to the above diagram, in each case above the moles are the same. The volume in the second is greater than the volume in the first, so the molarity in the second must be less than the molarity in the first.

If you take the equation separately for each solution:

$$moles\,(mol) = molarity\,(mol\;l^{-1})\times volume\,(l)$$

Moles (mol) are the same for each, and so the two equations can combine:

$$molarity\,(mol\;l^{-1})\times volume\,(l) = moles\,(mol) \qquad moles\,(mol) = molarity\,(mol\;l^{-1})\times volume\,(l)$$

This can be abreviated to:

$$molarity\,(mol\;l^{-1})\times volume\,(l) = molarity\,(mol\;l^{-1})\times volume\,(l)$$

This means the molarity multiplied by the volume of the first solution equals the molarity multiplied by the volume of the second equation (when the moles are the same).

$$M_1V_1 = M_2V_2$$

So, this equation allows you to work out the concentration of a solution before or after it is diluted, or the volume required to make a solution of a specific concentration, or molarity.

A note about units

On the whole it is best to work in grams and litres. If a question is in ml, remember that there are 1000 ml in one litre. It is best to convert units before you start answering the question.

Check you understand....

Answer the following questions:

1) What is the difference between moles, molar mass and molarity?
2) What are the units of molar mass?
3) What are the units of molarity?
4) If a solution is diluted what changes, the moles or the molarity?

For example: Your final one litre of solution must contain 1.0 mol l^{-1} $(NH_4)2SO_4$ (ammonium sulphate). How much stock 3.0mol l^{-1} $(NH_4)2SO_4$ must you add?

How to tackle a question like this

Identify what you have to work out.

The question asks 'How much stock must you add'. The question is asking what volume you must add to get a specific dilution.

Identify what you are given.

The final volume you require (1 litre),
The molarity of the solution you are given (3.0 mol l^{-1})
The molarity of the solution you want to make 1.0 mol l^{-1}

Work out the equation you need.

When you are taking one solution and making another, you will be using the equation:

$$M_1V_1 = M_2V_2$$

Rearrange the equation.

The question asks how much stock you must add, so it wants you to find V_1. The equation must therefore be rearranged to:

$$V_1 = \frac{M_2V_2}{M_1}$$

Plug in the numbers.

$$V_1 = \frac{1.0\,mol\,l^{-1} \times 1\,l}{3.0\,mol\,l^{-1}} = 0.33\,l$$

Check the answer.

Does the answer make sense? You will be adding 0.33 l and making it up to one litre. This means you are diluting you solution by three. Since your initial molarity was 3.0 mol l^{-1} and your final molarity is three times smaller. This answer appears sensible.

A more complicated example:

You are provided with NaCl as a 5 mol l^{-1} solution, KH_2PO_4 as a solid and glucose as a solution of 550g l^{-1}. Molar mass NaCl = 58.4 g mol^{-1}, molar mass KH_2PO_4 = 136.1 g mol^{-1}, molar mass glucose = 180.2 g mol^{-1}. You have been asked to make a one litre solution that contains a mole of each chemical. How much of each chemical would you need?

How to tackle a question like this

This question uses all the information you have been given in this chapter. You may panic and wish to scream, but it is not complicated. All you have to do is:

1. Split the question up.

2. For each chemical, work out what information you have and what you need.

3. Identify the equation or equations for each and plug in the numbers.

4. Look back and check the answers make sense.

NaCl is 5 mol l^{-1} solution. The molar mass of NaCl is 58.44 g mol^{-1}. Your final solution has a volume of one litre and contains 1 mol NaCl.

This goes from one solution to a second solution, and it is a dilution, so you need:

$$M_1V_1 = M_2V_2$$

Remember that molarity is number of moles per litre, so M_2 = 1 mol l^{-1}

$$V_1 = \frac{M_2V_2}{M_1} = \frac{1\,mol\,l^{-1} \times 1\,l}{5\,mol\,l^{-1}} = 0.2\,l$$

KH_2PO_4 is provided as a solid, so you will want to work out an answer that is in grams. You have been given the molar mass and you want to find the mass required to give a solution with 1 mole in one litre.

$$mass\,(g) = molarity\,(mol\,l^{-1}) \times molar\ mass\,(g\ mol^{-1}) \times volume\,(l)$$

$$mass\,(g) = 1\,mol\,l^{-1} \times 136.1\ g\ mol^{-1} \times 1\,l = 136.1\ g$$

For glucose you have been provided with a solution, and need to find out how much you must dilute it. However, rather than being given the molarity of the initial slution you are given the number of grams in one litre.

So first you use

$$molarity\,(mol\;l^{-1}) = \frac{mass(g)}{molar\;mass\,(g\;mol^{-1}) \times volume\,(l)} = \frac{550\;g}{180.2\;g\;mol^{-1} \times 1\,l} = 3.05\;mol\;l^{-1}$$

Then you can use

$$M_1V_1 = M_2V_2$$

You want to know the volume needed to make the dilution, so you need to rearrange the equation to give V1:

$$V_1 = \frac{M_2V_2}{M_1}$$

Final molarity = 1 mol l^{-1}

Final volume = 1 l

Initial molarity = 3.05 mol l^{-1}

$$V_1 = \frac{1.0\;mol\;l^{-1} \times 1\,l}{3.05\;mol\;l^{-1}} = 0.33\,l$$

So for your three chemicals you would need:

0.2 l of the NaCl solution

136.1 g KH_2PO_4

0.33 l of the glucose solution

If you were to physically make up this solution you would put the solid chemical and two solutions together, and then make up with water to the desired volume, in this case one litre.

Test your understanding.

1. How many moles of boric acid are there in 3 g boric acid? (Molar mass of boric acid = 62 g mol^{-1})

2. What is the molar mass of KOH if there are 4 moles in 224g of KOH?

3. If you had 0.5 g of KCl in 40 ml, what would the molarity of the KCl solution be? Molar mass of KCl is 74.5 g mol^{-1}.

4. Calculate the molarity of sodium carbonate in a 250 ml solution containing 10 g sodium carbonate. Molar mass of sodium carbonate is 106 g mol^{-1}.

5. How much 0.5 mol l^{-1} solution could you make from 10 ml of 5 mol l^{-1} stock?

6. If you had two stock solutions: 3 mol l^{-1} NaCl and 4 mol l^{-1} KOH, how much of each would you need to make a 30 ml solution of 0.4 mol l^{-1} NaCl and 0.6 mol l^{-1} KOH?

7. The molar mass of NaCl is 58.44 g mol^{-1}. How many moles of NaCl are there in 11.68 g?

8. What is the molarity of a 200 ml solution that contains 2.92 g NaCl, with a molar mass of 58.44 g mol^{-1}?

9. How much of a 1 mol l^{-1} solution of KOH would you need to make 200 ml of a 0.25 mol l^{-1} solution of KOH?

10. The molar mass of glucose is 180 g mol^{-1}. How many moles are there in 360 g glucose?

11. Given that there are 0.5 moles of KI in 83 g of KI, what is its molar mass?

12. The molar mass of boric acid is 62 g mol^{-1}. What mass of boric acid is required to give 1.5 moles?

13. A 500 ml solution contains 0.5844 g NaCl, with a molar mass of 58.44 g mol^{-1}. What is the molarity of this solution?

14. A 100 ml solution, with a molarity of 2 mol l^{-1} , was made using how much anhydrous sodium acetate, with a molar mass of 82 g mol^{-1}?

15. How much 0.25 mol l^{-1} solution can be made from 500 ml of a 1 mol l^{-1} NaCl stock?

16. You have been provided with a 1.2 l solution of NaCl which contains 0.8 moles of NaCl. How much of this would you need to create 300 ml of solution that has a concentration of 0.04 mol l^{-1} of NaCl?

Chapter 4
Fractions

03

Fractions

Objectives: This chapter:

- explains what is meant by a fraction
- defines the terms 'improper fraction', 'proper fraction' and 'mixed fraction'
- explains how to write fractions in different but equivalent forms
- explains how to simplify fractions by cancelling common factors
- explains how to add, subtract, multiply and divide fractions

Introduction

The arithmetic of fractions is very important groundwork that must be mastered before topics in algebra such as formulae and equations can be understood. The same techniques that are used to manipulate fractions are used in these more advanced topics. You should use this chapter to ensure that you are confident at handling fractions before moving on to algebra. In all the examples and exercises it is important that you should carry out the calculations without the use of a calculator.

Fractions are numbers such as $\frac{1}{2}$, $\frac{3}{4}$, $\frac{11}{8}$ and so on. In general a fraction is a number of the form $\frac{p}{q}$, where the letters p and q represent whole numbers or integers. The integer q can never be zero because it is never possible to divide by zero.

In any fraction $\frac{p}{q}$ the number p is called the **numerator** and the number q is called the **denominator**.

Key point

$$\text{fraction} = \frac{\text{numerator}}{\text{denominator}} = \frac{p}{q}$$

Suppose that p and q are both positive numbers. If p is less than q, the fraction is said to be a **proper fraction**. So $\frac{1}{2}$ and $\frac{3}{4}$ are proper fractions since the numerator is less than the denominator. If p is greater than or equal to q, the fraction is said to be **improper**. So $\frac{11}{8}$, $\frac{7}{4}$ and $\frac{3}{3}$ are all improper fractions.

If either of p or q is negative, we simply ignore the negative sign when determining whether the fraction is proper or improper. So $-\frac{3}{5}$, $\frac{-7}{21}$ and $\frac{4}{-21}$ are proper fractions, but $\frac{3}{-3}$, $\frac{-8}{2}$ and $-\frac{11}{2}$ are improper.

Note that all proper fractions have a value less than 1.

The denominator of a fraction can take the value 1, as in $\frac{3}{1}$ and $\frac{7}{1}$. In these cases the result is a whole number, 3 and 7.

Self-assessment questions 4.1

1. Explain the terms (a) fraction, (b) improper fraction, (c) proper fraction. In each case give an example of your own.
2. Explain the terms (a) numerator, (b) denominator.

Exercise 4.1

1. Classify each of the following as proper or improper:
   (a) $\frac{9}{17}$ (b) $\frac{-9}{17}$ (c) $\frac{8}{8}$ (d) $-\frac{7}{8}$ (e) $\frac{110}{77}$

Expressing a fraction in equivalent forms

Given a fraction, we may be able to express it in a different form. For example you will know that $\frac{1}{2}$ is equivalent to $\frac{2}{4}$. Note that multiplying both numerator and denominator by the same number leaves the value of the fraction unchanged. So, for example,

$$\frac{1}{2} = \frac{1 \times 2}{2 \times 2} = \frac{2}{4}$$

We say that $\frac{1}{2}$ and $\frac{2}{4}$ are **equivalent fractions**. Although they might look different, they have the same value.

Similarly, given the fraction $\frac{8}{12}$ we can divide both numerator and denominator by 4 to obtain

$$\frac{8}{12} = \frac{8/4}{12/4} = \frac{2}{3}$$

so $\frac{8}{12}$ and $\frac{2}{3}$ have the same value and are equivalent fractions.

Key point

Multiplying or dividing both numerator and denominator of a fraction by the same number produces a fraction having the same value, called an equivalent fraction.

A fraction is in its **simplest form** when there are no factors common to both numerator and denominator. For example $\frac{5}{12}$ is in its simplest form, but $\frac{3}{6}$ is not since 3 is a factor common to both numerator and denominator. Its simplest form is the equivalent fraction $\frac{1}{2}$.

To express a fraction in its simplest form we look for factors that are common to both the numerator and denominator. This is done by prime factorising both of these. Dividing both the numerator and denominator by any common factors removes them but leaves an equivalent fraction. This is equivalent to cancelling any common factors. For example, to simplify $\frac{4}{6}$ we prime factorise to produce

$$\frac{4}{6} = \frac{2 \times 2}{2 \times 3}$$

Dividing both numerator and denominator by 2 leaves $\frac{2}{3}$. This is equivalent to cancelling the common factor of 2.

WORKED EXAMPLES

4.1 Express $\frac{24}{36}$ in its simplest form.

Solution We seek factors common to both numerator and denominator. To do this we prime factorise 24 and 36.

Prime factorisation has been described in §1.3.

$$24 = 2 \times 2 \times 2 \times 3 \qquad 36 = 2 \times 2 \times 3 \times 3$$

The factors $2 \times 2 \times 3$ are common to both 24 and 36 and so these may be cancelled. Note that only common factors may be cancelled when simplifying a fraction. Hence

Finding the highest common factor (h.c.f.) of two numbers is detailed in §1.4.

$$\frac{24}{36} = \frac{\not{2} \times \not{2} \times 2 \times \not{3}}{\not{2} \times \not{2} \times \not{3} \times 3} = \frac{2}{3}$$

In its simplest form $\frac{24}{36}$ is $\frac{2}{3}$. In effect we have divided 24 and 36 by 12, which is their h.c.f.

4.2 Express $\frac{49}{21}$ in its simplest form.

Solution Prime factorising 49 and 21 gives

$$49 = 7 \times 7 \qquad 21 = 3 \times 7$$

Their h.c.f. is 7. Dividing 49 and 21 by 7 gives

$$\frac{49}{21} = \frac{7}{3}$$

Hence the simplest form of $\frac{49}{21}$ is $\frac{7}{3}$.

Before we can start to add and subtract fractions it is necessary to be able to convert fractions into a variety of equivalent forms. Work through the following examples.

WORKED EXAMPLES

4.3 Express $\frac{3}{4}$ as an equivalent fraction having a denominator of 20.

Solution To achieve a denominator of 20, the existing denominator must be multiplied by 5. To produce an equivalent fraction both numerator and denominator must be multiplied by 5, so

$$\frac{3}{4} = \frac{3 \times 5}{4 \times 5} = \frac{15}{20}$$

4.4 Express 7 as an equivalent fraction with a denominator of 3.

Solution Note that 7 is the same as the fraction $\frac{7}{1}$. To achieve a denominator of 3, the existing denominator must be multiplied by 3. To produce an equivalent fraction both numerator and denominator must be multiplied by 3, so

$$7 = \frac{7}{1} = \frac{7 \times 3}{1 \times 3} = \frac{21}{3}$$

Self-assessment questions 4.2

1. All integers can be thought of as fractions. True or false?
2. Explain the use of h.c.f. in the simplification of fractions.
3. Give an example of three fractions that are equivalent.

Exercise 4.2

1. Express the following fractions in their simplest form:
(a) $\frac{18}{27}$ (b) $\frac{12}{20}$ (c) $\frac{15}{45}$ (d) $\frac{25}{80}$ (e) $\frac{15}{60}$
(f) $\frac{90}{200}$ (g) $\frac{15}{20}$ (h) $\frac{2}{18}$ (i) $\frac{16}{24}$ (j) $\frac{30}{65}$
(k) $\frac{12}{21}$ (l) $\frac{100}{45}$ (m) $\frac{6}{9}$ (n) $\frac{12}{16}$ (o) $\frac{13}{42}$
(p) $\frac{13}{39}$ (q) $\frac{11}{33}$ (r) $\frac{14}{30}$ (s) $-\frac{12}{16}$ (t) $\frac{11}{-33}$
(u) $\frac{-14}{-30}$

2. Express $\frac{3}{4}$ as an equivalent fraction having a denominator 28.

3. Express 4 as an equivalent fraction with a denominator of 5.

4. Express $\frac{5}{12}$ as an equivalent fraction having a denominator of 36.

5. Express 2 as an equivalent fraction with a denominator of 4.

6. Express 6 as an equivalent fraction with a denominator of 3.

7. Express each of the fractions $\frac{2}{3}$, $\frac{5}{4}$ and $\frac{5}{6}$ as an equivalent fraction with a denominator of 12.

8. Express each of the fractions $\frac{4}{9}$, $\frac{1}{2}$ and $\frac{5}{6}$ as an equivalent fraction with a denominator of 18.

9. Express each of the following numbers as an equivalent fraction with a denominator of 12:
(a) $\frac{1}{2}$ (b) $\frac{3}{4}$ (c) $\frac{5}{2}$ (d) 5 (e) 4 (f) 12

Addition and subtraction of fractions

To add and subtract fractions we first rewrite each fraction so that they all have the same denominator. This is known as the **common denominator**. The denominator is chosen to be the lowest common multiple of the original denominators. Then the numerators only are added or subtracted as appropriate, and the result is divided by the common denominator.

WORKED EXAMPLES

4.5 Find $\frac{2}{3}+\frac{5}{4}$.

Solution The denominators are 3 and 4. The l.c.m. of 3 and 4 is 12. We need to express both fractions with a denominator of 12.

Finding the lowest common multiple (l.c.m.) is detailed in §1.4.

To express $\frac{2}{3}$ with a denominator of 12 we multiply both numerator and denominator by 4. Hence $\frac{2}{3}$ is the same as $\frac{8}{12}$. To express $\frac{5}{4}$ with a denominator of 12 we multiply both numerator and denominator by 3. Hence $\frac{5}{4}$ is the same as $\frac{15}{12}$. So

$$\frac{2}{3}+\frac{5}{4}=\frac{8}{12}+\frac{15}{12}=\frac{8+15}{12}=\frac{23}{12}$$

4.6 Find $\frac{4}{9} - \frac{1}{2} + \frac{5}{6}$.

Solution The denominators are 9, 2 and 6. Their l.c.m. is 18. Each fraction is expressed with 18 as the denominator:

$$\frac{4}{9} = \frac{8}{18} \qquad \frac{1}{2} = \frac{9}{18} \qquad \frac{5}{6} = \frac{15}{18}$$

Then

$$\frac{4}{9} - \frac{1}{2} + \frac{5}{6} = \frac{8}{18} - \frac{9}{18} + \frac{15}{18} = \frac{8 - 9 + 15}{18} = \frac{14}{18}$$

The fraction $\frac{14}{18}$ can be simplified to $\frac{7}{9}$. Hence

$$\frac{4}{9} - \frac{1}{2} + \frac{5}{6} = \frac{7}{9}$$

4.7 Find $\frac{1}{4} - \frac{5}{9}$.

Solution The l.c.m. of 4 and 9 is 36. Each fraction is expressed with a denominator of 36. Thus

$$\frac{1}{4} = \frac{9}{36} \quad \text{and} \quad \frac{5}{9} = \frac{20}{36}$$

Then

$$\frac{1}{4} - \frac{5}{9} = \frac{9}{36} - \frac{20}{36}$$

$$= \frac{9 - 20}{36}$$

$$= \frac{-11}{36}$$

$$= -\frac{11}{36}$$

Consider the number $2\frac{3}{4}$. This is referred to as a **mixed fraction** because it contains a whole number part, 2, and a fractional part, $\frac{3}{4}$. We can convert this mixed fraction into an improper fraction as follows. Recognise that 2 is equivalent to $\frac{8}{4}$, and so $2\frac{3}{4}$ is $\frac{8}{4} + \frac{3}{4} = \frac{11}{4}$.

The reverse of this process is to convert an improper fraction into a mixed fraction. Consider the improper fraction $\frac{11}{4}$. Now 4 divides into 11 twice leaving a remainder of 3; so $\frac{11}{4} = 2$ remainder 3, which we write as $2\frac{3}{4}$.

WORKED EXAMPLE

4.8 (a) Express $4\frac{2}{5}$ as an improper fraction.

(b) Find $4\frac{2}{5} + \frac{1}{3}$.

Solution (a) $4\frac{2}{5}$ is a mixed fraction. Note that $4\frac{2}{5}$ is equal to $4 + \frac{2}{5}$. We can write 4 as the equivalent fraction $\frac{20}{5}$. Therefore

$$4\frac{2}{5} = \frac{20}{5} + \frac{2}{5}$$

$$= \frac{22}{5}$$

(b) $$4\frac{2}{5} + \frac{1}{3} = \frac{22}{5} + \frac{1}{3}$$

$$= \frac{66}{15} + \frac{5}{15}$$

$$= \frac{71}{15}$$

Self-assessment question 4.3

1. Explain the use of l.c.m. when adding and subtracting fractions.

Exercise 4.3

1. Find

(a) $\frac{1}{4} + \frac{2}{3}$ (b) $\frac{3}{5} + \frac{5}{3}$ (c) $\frac{12}{14} - \frac{2}{7}$

(d) $\frac{3}{7} - \frac{1}{2} + \frac{2}{21}$ (e) $1\frac{1}{2} + \frac{4}{9}$

(f) $2\frac{1}{4} - 1\frac{1}{3} + \frac{1}{2}$ (g) $\frac{10}{15} - 1\frac{2}{5} + \frac{8}{3}$

(h) $\frac{9}{10} - \frac{7}{16} + \frac{1}{2} - \frac{2}{5}$

2. Find

(a) $\frac{7}{8} + \frac{1}{3}$ (b) $\frac{1}{2} - \frac{3}{4}$ (c) $\frac{3}{5} + \frac{2}{3} + \frac{1}{2}$

(d) $\frac{3}{8} + \frac{1}{3} + \frac{1}{4}$ (e) $\frac{2}{3} - \frac{4}{7}$

(f) $\frac{1}{11} - \frac{1}{2}$ (g) $\frac{3}{11} - \frac{5}{8}$

3. Express as improper fractions:
(a) $2\frac{1}{2}$ (b) $3\frac{2}{3}$ (c) $10\frac{1}{4}$ (d) $5\frac{2}{7}$
(e) $6\frac{2}{9}$ (f) $11\frac{1}{3}$ (g) $15\frac{1}{2}$ (h) $13\frac{3}{4}$
(i) $12\frac{1}{11}$ (j) $13\frac{2}{3}$ (k) $56\frac{1}{2}$

4. Without using a calculator express these improper fractions as mixed fractions:
(a) $\frac{10}{3}$ (b) $\frac{7}{2}$ (c) $\frac{15}{4}$ (d) $\frac{25}{6}$

Multiplication of fractions

The product of two or more fractions is found by multiplying their numerators to form a new numerator, and then multiplying their denominators to form a new denominator.

WORKED EXAMPLES

4.9 Find $\frac{4}{9} \times \frac{3}{8}$.

Solution The numerators are multiplied: $4 \times 3 = 12$. The denominators are multiplied: $9 \times 8 = 72$. Hence

$$\frac{4}{9} \times \frac{3}{8} = \frac{12}{72}$$

This may now be expressed in its simplest form:

$$\frac{12}{72} = \frac{1}{6}$$

Hence

$$\frac{4}{9} \times \frac{3}{8} = \frac{1}{6}$$

An alternative, but equivalent, method is to cancel any factors common to both numerator and denominator at the outset:

$$\frac{4}{9} \times \frac{3}{8} = \frac{4 \times 3}{9 \times 8}$$

A factor of 4 is common to the 4 and the 8. Hence

$$\frac{4 \times 3}{9 \times 8} = \frac{1 \times 3}{9 \times 2}$$

A factor of 3 is common to the 3 and the 9. Hence

$$\frac{1 \times 3}{9 \times 2} = \frac{1 \times 1}{3 \times 2} = \frac{1}{6}$$

4.10 Find $\frac{12}{25} \times \frac{2}{7} \times \frac{10}{9}$.

Solution We cancel factors common to both numerator and denominator. A factor of 5 is common to 10 and 25. Cancelling this gives

$$\frac{12}{25} \times \frac{2}{7} \times \frac{10}{9} = \frac{12}{5} \times \frac{2}{7} \times \frac{2}{9}$$

A factor of 3 is common to 12 and 9. Cancelling this gives

$$\frac{12}{5} \times \frac{2}{7} \times \frac{2}{9} = \frac{4}{5} \times \frac{2}{7} \times \frac{2}{3}$$

There are no more common factors. Hence

$$\frac{12}{25} \times \frac{2}{7} \times \frac{10}{9} = \frac{4}{5} \times \frac{2}{7} \times \frac{2}{3} = \frac{16}{105}$$

4.11 Find $\frac{3}{4}$ of $\frac{5}{9}$.

Recall that 'of' means multiply

Solution $\frac{3}{4}$ of $\frac{5}{9}$ is the same as $\frac{3}{4} \times \frac{5}{9}$. Cancelling a factor of 3 from numerator and denominator gives $\frac{1}{4} \times \frac{5}{3}$, that is $\frac{5}{12}$. Hence $\frac{3}{4}$ of $\frac{5}{9}$ is $\frac{5}{12}$.

4.12 Find $\frac{5}{6}$ of 70.

Solution We can write 70 as $\frac{70}{1}$. So

$$\frac{5}{6} \text{ of } 70 = \frac{5}{6} \times \frac{70}{1} = \frac{5}{3} \times \frac{35}{1} = \frac{175}{3} = 58\,\frac{1}{3}$$

4.13 Find $2\frac{7}{8} \times \frac{2}{3}$.

Solution In this example the first fraction is a mixed fraction. We convert it to an improper fraction before performing the multiplication. Note that $2\frac{7}{8} = \frac{23}{8}$. Then

$$\frac{23}{8} \times \frac{2}{3} = \frac{23}{4} \times \frac{1}{3}$$

$$= \frac{23}{12}$$

$$= 1\frac{11}{12}$$

Self-assessment question 4.4

1. Describe how to multiply fractions together.

Exercise 4.4

1. Evaluate

   (a) $\frac{2}{3} \times \frac{6}{7}$ (b) $\frac{8}{15} \times \frac{25}{32}$ (c) $\frac{1}{4} \times \frac{8}{9}$

   (d) $\frac{16}{17} \times \frac{34}{48}$ (e) $2 \times \frac{3}{5} \times \frac{5}{12}$

   (f) $2\frac{1}{3} \times 1\frac{1}{4}$ (g) $1\frac{3}{4} \times 2\frac{1}{2}$

   (h) $\frac{3}{4} \times 1\frac{1}{2} \times 3\frac{1}{2}$

2. Evaluate

   (a) $\frac{2}{3}$ of $\frac{3}{4}$ (b) $\frac{4}{7}$ of $\frac{21}{30}$

   (c) $\frac{9}{10}$ of 80 (d) $\frac{6}{7}$ of 42

3. Is $\frac{3}{4}$ of $\frac{12}{15}$ the same as $\frac{12}{15}$ of $\frac{3}{4}$?

4. Find

   (a) $-\frac{1}{3} \times \frac{5}{7}$ (b) $\frac{3}{4} \times -\frac{1}{2}$

   (c) $\left(-\frac{5}{8}\right) \times \frac{8}{11}$ (d) $\left(-\frac{2}{3}\right) \times \left(-\frac{15}{7}\right)$

5. Find

   (a) $5\frac{1}{2} \times \frac{1}{2}$ (b) $3\frac{3}{4} \times \frac{1}{3}$

   (c) $\frac{2}{3} \times 5\frac{1}{9}$ (d) $\frac{3}{4} \times 11\frac{1}{2}$

6. Find

   (a) $\frac{3}{5}$ of $11\frac{1}{4}$ (b) $\frac{2}{3}$ of $15\frac{1}{2}$

   (c) $\frac{1}{4}$ of $-8\frac{1}{3}$

Division by a fraction

To divide one fraction by another fraction, we invert the second fraction and then multiply. When we invert a fraction we interchange the numerator and denominator.

WORKED EXAMPLES

4.14 Find $\frac{6}{25} \div \frac{2}{5}$.

Solution We invert $\frac{2}{5}$ to obtain $\frac{5}{2}$. Multiplication is then performed. So

$$\frac{6}{25} \div \frac{2}{5} = \frac{6}{25} \times \frac{5}{2} = \frac{3}{25} \times \frac{5}{1} = \frac{3}{5} \times \frac{1}{1} = \frac{3}{5}$$

4.15 Evaluate (a) $1\frac{1}{3} \div \frac{8}{3}$ (b) $\frac{20}{21} \div \frac{5}{7}$.

Solution (a) First we express $1\frac{1}{3}$ as an improper fraction:

$$1\frac{1}{3} = 1 + \frac{1}{3} = \frac{3}{3} + \frac{1}{3} = \frac{4}{3}$$

So we calculate

$$\frac{4}{3} \div \frac{8}{3} = \frac{4}{3} \times \frac{3}{8} = \frac{4}{8} = \frac{1}{2}$$

Hence

$$1\frac{1}{3} \div \frac{8}{3} = \frac{1}{2}$$

(b) $$\frac{20}{21} \div \frac{5}{7} = \frac{20}{21} \times \frac{7}{5} = \frac{4}{21} \times \frac{7}{1} = \frac{4}{3}$$

Self-assessment question 4.5

1. Explain the process of division by a fraction.

Exercise 4.5

1. Evaluate

(a) $\frac{3}{4} \div \frac{1}{8}$ (b) $\frac{8}{9} \div \frac{4}{3}$ (c) $\frac{-2}{7} \div \frac{4}{21}$ (d) $\frac{9}{4} \div 1\frac{1}{2}$

(e) $\frac{5}{6} \div \frac{5}{12}$ (f) $\frac{99}{100} \div 1\frac{4}{5}$ (g) $3\frac{1}{4} \div 1\frac{1}{8}$ (h) $\left(2\frac{1}{4} \div \frac{3}{4}\right) \times 2$

(i) $2\frac{1}{4} \div \left(\frac{3}{4} \times 2\right)$ (j) $6\frac{1}{4} \div 2\frac{1}{2} + 5$ (k) $6\frac{1}{4} \div \left(2\frac{1}{2} + 5\right)$

Test and assignment exercises 4.1

1. Evaluate

(a) $\frac{3}{4} + \frac{1}{6}$ (b) $\frac{2}{3} + \frac{3}{5} - \frac{1}{6}$ (c) $\frac{5}{7} - \frac{2}{3}$ (d) $2\frac{1}{3} - \frac{9}{10}$

(e) $5\frac{1}{4} + 3\frac{1}{6}$ (f) $\frac{9}{8} - \frac{7}{6} + 1$ (g) $\frac{5}{6} - \frac{5}{3} + \frac{5}{4}$ (h) $\frac{4}{5} + \frac{1}{3} - \frac{3}{4}$

2. Evaluate

(a) $\frac{4}{7} \times \frac{21}{32}$ (b) $\frac{5}{6} \times \frac{8}{15}$ (c) $\frac{3}{11} \times \frac{20}{21}$ (d) $\frac{9}{14} \times \frac{8}{18}$

(e) $\frac{5}{4} \div \frac{10}{13}$ (f) $\frac{7}{16} \div \frac{21}{32}$ (g) $\frac{-24}{25} \div \frac{51}{50}$ (h) $\frac{45}{81} \div \frac{25}{27}$

3. Evaluate the following expressions using the BODMAS rule:

(a) $\frac{1}{2} + \frac{1}{3} \times 2$ (b) $\frac{3}{4} \times \frac{2}{3} + \frac{1}{4}$ (c) $\frac{5}{6} \div \frac{2}{3} + \frac{3}{4}$ (d) $\left(\frac{2}{3} + \frac{1}{4}\right) \div 4 + \frac{3}{5}$

(e) $\left(\frac{4}{3} - \frac{2}{5} \times \frac{1}{3}\right) \times \frac{1}{4} + \frac{1}{2}$ (f) $\frac{3}{4}$ of $\left(1 + \frac{2}{3}\right)$ (g) $\frac{2}{3}$ of $\frac{1}{2} + 1$ (h) $\frac{1}{5} \times \frac{2}{3} + \frac{2}{5} \div \frac{4}{5}$

4. Express in their simplest form:

(a) $\frac{21}{84}$ (b) $\frac{6}{80}$ (c) $\frac{34}{85}$ (d) $\frac{22}{143}$ (e) $\frac{69}{253}$

03

Decimal fractions

Objectives: This chapter:

- revises the decimal number system
- shows how to write a number to a given number of significant figures
- shows how to write a number to a given number of decimal places

Decimal numbers

Consider the whole number 478. We can regard it as the sum

$$400 + 70 + 8$$

In this way we see that, in the number 478, the 8 represents 8 ones, or 8 units, the 7 represents 7 tens, or 70, and the number 4 represents 4 hundreds or 400. Thus we have the system of hundreds, tens and units familiar from early years in school. All whole numbers can be thought of in this way.

When we wish to deal with proper fractions and mixed fractions, we extend the hundreds, tens and units system as follows. A **decimal point**, '.', marks the end of the whole number part, and the numbers that follow it, to the right, form the fractional part.

A number immediately to the right of the decimal point, that is, in the **first decimal place**, represents tenths, so

$$0.1 = \frac{1}{10}$$

$$0.2 = \frac{2}{10} \quad \text{or} \quad \frac{1}{5}$$

$$0.3 = \frac{3}{10} \quad \text{and so on}$$

Note that when there are no whole numbers involved it is usual to write a zero in front of the decimal point, thus, .2 would be written 0.2.

WORKED EXAMPLE

4.16 Express the following decimal numbers as proper fractions in their simplest form

(a) 0.4 (b) 0.5 (c) 0.6

Solution The first number after the decimal point represents tenths.

(a) $0.4 = \frac{4}{10}$, which simplifies to $\frac{2}{5}$

(b) $0.5 = \frac{5}{10}$ or simply $\frac{1}{2}$

(c) $0.6 = \frac{6}{10} = \frac{3}{5}$

Frequently we will deal with numbers having a whole number part and a fractional part. Thus

$$\begin{aligned} 5.2 &= 5 \text{ units} + 2 \text{ tenths} \\ &= 5 + \frac{2}{10} \\ &= 5 + \frac{1}{5} \\ &= 5\frac{1}{5} \end{aligned}$$

Similarly,

$$175.8 = 175\frac{8}{10} = 175\frac{4}{5}$$

Numbers in the second position after the decimal point, or the **second decimal place**, represent hundredths, so

$$0.01 = \frac{1}{100}$$

$$0.02 = \frac{2}{100} \quad \text{or} \quad \frac{1}{50}$$

$$0.03 = \frac{3}{100} \quad \text{and so on}$$

Consider 0.25. We can think of this as

$$0.25 = 0.2 + 0.05$$
$$= \frac{2}{10} + \frac{5}{100}$$
$$= \frac{25}{100}$$

We see that 0.25 is equivalent to $\frac{25}{100}$, which in its simplest form is $\frac{1}{4}$.

In fact we can regard any numbers occupying the first two decimal places as hundredths, so that

$$0.25 = \frac{25}{100} \quad \text{or simply } \frac{1}{4}$$

$$0.50 = \frac{50}{100} \quad \text{or } \frac{1}{2}$$

$$0.75 = \frac{75}{100} = \frac{3}{4}$$

WORKED EXAMPLES

4.17 Express the following decimal numbers as proper fractions in their simplest form:

(a) 0.35 (b) 0.56 (c) 0.68

Solution The first two decimal places represent hundredths:

(a) $0.35 = \frac{35}{100} = \frac{7}{20}$

(b) $0.56 = \frac{56}{100} = \frac{14}{25}$

(c) $0.68 = \frac{68}{100} = \frac{17}{25}$

4.18 Express 37.25 as a mixed fraction in its simplest form.

Solution

$$37.25 = 37 + 0.25$$
$$= 37 + \frac{25}{100}$$
$$= 37 + \frac{1}{4}$$
$$= 37\frac{1}{4}$$

Numbers in the third position after the decimal point, or **third decimal place**, represent thousandths, so

$$0.001 = \frac{1}{1000}$$

$$0.002 = \frac{2}{1000} \quad \text{or} \quad \frac{1}{500}$$

$$0.003 = \frac{3}{1000} \quad \text{and so on}$$

In fact we can regard any numbers occupying the first three positions after the decimal point as thousandths, so that

$$0.356 = \frac{356}{1000} \quad \text{or} \quad \frac{89}{250}$$

$$0.015 = \frac{15}{1000} \quad \text{or} \quad \frac{3}{200}$$

$$0.075 = \frac{75}{1000} = \frac{3}{40}$$

WORKED EXAMPLE

4.19 Write each of the following as a decimal number:

(a) $\frac{3}{10} + \frac{7}{100}$ (b) $\frac{8}{10} + \frac{3}{1000}$

Solution

(a) $\frac{3}{10} + \frac{7}{100} = 0.3 + 0.07 = 0.37$

(b) $\frac{8}{10} + \frac{3}{1000} = 0.8 + 0.003 = 0.803$

You will normally use a calculator to add, subtract, multiply and divide decimal numbers. Generally the more decimal places used, the more accurately we can state a number. This idea is developed in the next section.

Self-assessment questions 4.6

1. State which is the largest and which is the smallest of the following numbers: 23.001, 23.0, 23.00001, 23.0008, 23.01

2. Which is the largest of the following numbers? 0.1, 0.02, 0.003, 0.0004, 0.00005

Exercise 4.6

1. Express the following decimal numbers as proper fractions in their simplest form:
(a) 0.7 (b) 0.8 (c) 0.9

2. Express the following decimal numbers as proper fractions in their simplest form:
(a) 0.55 (b) 0.158 (c) 0.98
(d) 0.099

3. Express each of the following as a mixed fraction in its simplest form:
(a) 4.6 (b) 5.2 (c) 8.05 (d) 11.59
(e) 121.09

4. Write each of the following as a decimal number:
(a) $\frac{6}{10} + \frac{9}{100} + \frac{7}{1000}$ (b) $\frac{8}{100} + \frac{3}{1000}$
(c) $\frac{17}{1000} + \frac{5}{10}$

Significant figures and decimal places

The accuracy with which we state a number often depends upon the context in which the number is being used. The volume of a petrol tank is usually given to the nearest litre. It is of no practical use to give such a volume to the nearest cubic centimetre.

When writing a number we often give the accuracy by stating the **number of significant figures** or the **number of decimal places** used. These terms are now explained.

Significant figures

Suppose we are asked to write down the number nearest to 857 using at most two non-zero digits, or numbers. We would write 860. This number is nearer to 857 than any other number with two non-zero digits. We say that 857 to 2 **significant figures** is 860. The words 'significant figures' are usually abbreviated to s.f. Because 860 is larger than 857 we say that the 857 has been **rounded up** to 860.

To write a number to three significant figures we can use no more than three non-zero digits. For example, the number closest to 1784 which has no more than three non-zero digits is 1780. We say that 1784 to 3 significant figures is 1780. In this case, because 1780 is less than 1784 we say that 1784 has been **rounded down** to 1780.

WORKED EXAMPLES

4.20 Write down the number nearest to 86 using only one non-zero digit. Has 86 been rounded up or down?

Solution The number 86 written to one significant figure is 90. This number is nearer to 86 than any other number having only one non-zero digit. 86 has been rounded up to 90.

4.21 Write down the number nearest to 999 which uses only one non-zero digit.

Solution The number 999 to one significant figure is 1000. This number is nearer to 999 than any other number having only one non-zero digit.

We now explain the process of writing to a given number of significant figures.

When asked to write a number to, say, three significant figures, 3 s.f., the first step is to look at the first four digits. If asked to write a number to two significant figures we look at the first three digits and so on. We always look at one more digit than the number of significant figures required.

For example, to write 6543.19 to 2 s.f. we would consider the number 6540.00; the digits 3, 1 and 9 are effectively ignored. The next step is to round up or down. If the final digit is a 5 or more then we round up by increasing the previous digit by 1. If the final digit is 4 or less we round down by leaving the previous digit unchanged. Hence when considering 6543.19 to 2 s.f., the 4 in the third place means that we round down to 6500.

To write 23865 to 3 s.f. we would consider the number 23860. The next step is to increase the 8 to a 9. Thus 23865 is rounded up to 23900.

Zeros at the beginning of a number are ignored. To write 0.004693 to 2 s.f. we would first consider the number 0.00469. Note that the zeros at the beginning of the number have not been counted. We then round the 6 to a 7, producing 0.0047.

The following examples illustrate the process.

WORKED EXAMPLES

4.22 Write 36.482 to 3 s.f.

Solution We consider the first 4 digits, that is, 36.48. The final digit is 8 and so we round up 36.48 to 36.5. To 3 s.f. 36.482 is 36.5.

4.23 Write 1.0049 to 4 s.f.

Solution To write to 4 s.f. we consider the first 5 digits, that is, 1.0049. The final digit is a 9 and so 1.0049 is rounded up to 1.005.

4.24 Write 695.3 to 2 s.f.

Solution We consider 695. The final digit is a 5 and so we round up. We cannot round up the 9 to a 10 and so the 69 is rounded up to 70. Hence to 2 s.f. the number is 700.

4.25 Write 0.0473 to 1 s.f.

Solution We do not count the initial zeros and consider 0.047. The final digit tells us to round up. Hence to 1 s.f. we have 0.05.

4.26 A number is given to 2 s.f. as 67.

(a) What is the maximum value the number could have?
(b) What is the minimum value the number could have?

Solution

(a) To 2 s.f. 67.5 is 68. Any number just below 67.5, for example, 67.49 or 67.499, to 2 s.f. is 67. Hence the maximum value of the number is 67.4999....

(b) To 2 s.f. 66.4999... is 66. However, 66.5 to 2 s.f. is 67. The minimum value of the number is thus 66.5.

Decimal places

When asked to write a number to 3 decimal places (3 d.p.) we consider the first 4 decimal places, that is, numbers after the decimal point. If asked to write to 2 d.p. we consider the first 3 decimal places and so on. If the final digit is 5 or more we round up, otherwise we round down.

WORKED EXAMPLES

4.27 Write 63.4261 to 2 d.p.

Solution We consider the number to 3 d.p., that is, 63.426. The final digit is 6 and so we round up 63.426 to 63.43. Hence 63.4261 to 2 d.p. is 63.43.

4.28 Write 1.97 to 1 d.p.

Solution In order to write to 1 d.p. we consider the number to 2 d.p., that is, we consider 1.97. The final digit is a 7 and so we round up. The 9 cannot be rounded up and so we look at 1.9. This can be rounded up to 2.0. Hence 1.97 to 1 d.p. is 2.0. Note that it is crucial to write 2.0 and not simply 2, as this shows that the number is written to 1 d.p.

4.29 Write −6.0439 to 2 d.p.

Solution We consider −6.043. As the final digit is a 3 the number is rounded down to −6.04.

Self-assessment questions 4.7

1. Explain the meaning of 'significant figures'.
2. Explain the process of writing a number to so many decimal places.

Exercise 4.7

1. Write to 3 s.f.
   (a) 6962 (b) 70.406 (c) 0.0123
   (d) 0.010991 (e) 45.607 (f) 2345
2. Write 65.999 to
   (a) 4 s.f. (b) 3 s.f. (c) 2 s.f.
   (d) 1 s.f. (e) 2 d.p. (f) 1 d.p.
3. Write 9.99 to
   (a) 1 s.f. (b) 1 d.p.
4. Write 65.4555 to
   (a) 3 d.p. (b) 2 d.p. (c) 1 d.p.
   (d) 5 s.f. (e) 4 s.f. (f) 3 s.f. (g) 2 s.f.
   (h) 1 s.f.

Test and assignment exercises 4.2

1. Express the following numbers as proper fractions in their simplest form:
   (a) 0.74 (b) 0.96 (c) 0.05 (d) 0.25
2. Express each of the following as a mixed fraction in its simplest form:
   (a) 2.5 (b) 3.25 (c) 3.125 (d) 6.875
3. Write each of the following as a decimal number:
   (a) $\frac{3}{10}+\frac{1}{100}+\frac{7}{1000}$ (b) $\frac{5}{1000}+\frac{9}{100}$ (c) $\frac{4}{1000}+\frac{9}{10}$
4. Write 0.09846 to (a) 1 d.p, (b) 2 s.f., (c) 1 s.f.
5. Write 9.513 to (a) 3 s.f., (b) 2 s.f., (c) 1 s.f.
6. Write 19.96 to (a) 1 d.p., (b) 2 s.f., (c) 1 s.f.

Applying decimal places

Deciding how many decimal places and significant figures to use

When you carry out a calculation using a calculator it gives you the answer to many significant figures. While carrying out a calculation it is a good idea to use many significant figures. However, it is important to give the final answer with the number of significant figures that reflects the accuracy of the initial data.

For example, you need to work out the volume of a spherical cell. You measured the diameter and found it to be 6.3 μm. This means that you know its diameter to an accuracy of 0.1 μm; you know it is not 6.2 μm, nor 6.4 μm. It could be 6.34 μm, but the accuracy of your measuring does not provide you with a figure to 0.01μm.

The radius is half the diameter, so is 3.15 μm. It is okay to use more significant figures at this stage because it is the middle of the calculation.

Volume of a sphere is

$$V = \frac{4}{3}\pi r^3$$

So:

$$V = \frac{4}{3} \times 3.142 \times 3.15^3$$

$$V = 130.94127 \mu m^3$$

You should not give this as your final answer. It shows that you have not thought about your answer and what it means. You have just copied the numbers from your calculator.

Your original data was accurate to 0.1 μm. Your answer must reflect this, and so be given to one decimal place:

$$V = 130.9 \mu m^3$$

In this way you are communicating that you know the volume of the cell is not 131.0 μm nor 130.8 μm, and this is as accurate as you can be, given the initial measurement. The original answer to five decimal places gives the impression you can be accurate to ten pico meters, and this is clearly not true.

What happens if you round up a number in the middle of a calculation?
Taking the same example, if you calculated the radius of the cell as half the diameter, and rounded 3.15 μm to 3.2 μm before calculating the volume, the final answer would be:

$$V = 102.95705 \mu m^3$$

$$V = 103.0 \mu m^3$$

Here you can see the importance of only rounding up or down at the end of calculation, because the answer is very different if you round up or down in the middle of the calculation. This second answer would be wrong.

Test your understanding.

Work out the following, rounding up or down at the right time and giving the answer to an appropriate number of significant figures.

1. Calculate the speed that a bee flies after you have measured it covering 1.2 m in 3 s.

$$speed = \frac{distance}{time}$$

2. Calculate the area of a rectangular experimental plot where you have measured the length as 4.60 m and the width as 2.34 m.

3. Calculate the volume of an apple, assuming it to be spherical. You have measured the circumference to be 10.2 cm.

$$C = 2\pi r$$

$$V = \frac{4}{3}\pi r^3$$

4. You have counted 45 cells in an area 12 μm^2. How many would you expect to find in an area of 250 μm^2.

5. You have measured the radius of a cylinder to be 4.23 cm and the height to be 5.36 cm. Calculate the area, circumference and volume of the cylinder.

6. Calculate your running speed if you were to run around a circular track, that has a radius of 32 m, in 31 s.

7. If a tree grew at a steady rate of 1.3 m $year^{-1}$, and had a height of 4 m when you planted it, how tall would it be after 5 months? How tall would it be after 9 years?

8. A patient had a rash that covered 12 cm^2 of their back and was 1.7 times bigger after 3 days. Assuming it was growing at a steady rate, how many cm^2 was it increasing by each day?

9. The equation for body mass index (bmi) is $bmi = \frac{weight(kg)}{height(m)^2}$

If a patient had a height of 1.8 m and a weight of 74 kg, calculate their body mass index.

Chapter 5
Percentage and Ratio

03

Percentage and ratio

Objectives: This chapter:

- explains the terms 'percentage' and 'ratio'
- shows how to perform calculations using percentages and ratios

Percentage

A **percentage** is a fraction whose denominator is 100. We use the percent symbol %, to represent a percentage: for example, $\frac{17}{100}$ may be written as 17%.

WORKED EXAMPLE

5.1 Express $\frac{19}{100}$ and $\frac{35}{100}$ as percentages.

Solution Both of these fractions have a denominator of 100. So we can write simply

$$\frac{19}{100} = 19\%, \qquad \text{and} \qquad \frac{35}{100} = 35\%$$

Sometimes it is necessary to convert a fraction whose denominator is not 100 into a percentage. This can be done in two ways. The first way is to express the fraction as an equivalent fraction having a denominator of 100. Thus the fraction $\frac{2}{5}$ can be written as

$$\frac{2}{5} = \frac{2 \times 20}{5 \times 20} = \frac{40}{100}$$

$\frac{2}{5}$ and $\frac{40}{100}$ are equivalent fractions. We see that $\frac{2}{5} = 40\%$.

The second way in which a fraction can be converted into a percentage is to multiply the fraction by 100 and then label the result as a percentage. Thus to convert $\frac{2}{5}$ we write

$$\frac{2}{5} \times 100 = \frac{200}{5} = 40$$

and so $\frac{2}{5} = 40\%$.

Key point

To convert a fraction into a percentage, multiply by 100 and label the result as a percentage.

WORKED EXAMPLES

5.2 Convert $\frac{5}{8}$ into a percentage.

Solution We multiply $\frac{5}{8}$ by 100 to obtain

$$\frac{5}{8} \times 100 = \frac{500}{8} = 62.5$$

The result is labelled as a percentage. Thus $\frac{5}{8} = 62.5\%$.

5.3 Bill scores $\frac{13}{17}$ in a test. Mary scores $\frac{14}{19}$. Express the scores as percentages.

Solution To convert $\frac{13}{17}$ into a percentage we multiply by 100 and label the result as a percentage. Thus

$$\frac{13}{17} \times 100 = \frac{1300}{17} = 76.5 \text{ (1d.p.)}$$

So $\frac{13}{17} = 76.5\%$.
Similarly,

$$\frac{14}{19} \times 100 = \frac{1400}{19} = 73.7 \text{ (1d.p.)}$$

So $\frac{14}{19} = 73.7\%$.
In this form it is easier to make a comparison of the two marks.

5.4 Express 30% as a decimal.

Solution

$$30\% = \frac{30}{100} = 0.3$$

Hence 30% is equivalent to 0.3.

Some common percentage–fraction conversions are given here for reference.

Key point

$10\% = \frac{1}{10}$, $25\% = \frac{1}{4}$, $50\% = \frac{1}{2}$, $75\% = \frac{3}{4}$

Recall from §1.2 that 'of' means multiply.

We are often asked to calculate a percentage of a quantity. For example, find 50% of 160, or find 10% of 95. Such calculations frequently arise when calculating discounts on prices etc. Since $50\% = \frac{1}{2}$ we find

$$50\% \text{ of } 160 = \frac{1}{2} \times 160 = 80$$

and since $10\% = \frac{1}{10}$ we may write

$$10\% \text{ of } 95 = \frac{1}{10} \times 95 = 9.5$$

WORKED EXAMPLES

5.5 Calculate 27% of 90.

Solution Recall that 'of' means multiply.

$$27\% \text{ of } 90 = \frac{27}{100} \times 90 = \frac{27}{10} \times 9 = 24.3$$

5.6 Calculate 100% of 6.

Solution

$$100\% \text{ of } 6 = \frac{100}{100} \times 6 = 1 \times 6 = 6$$

From Worked Example 5.6 we note that 100% of a number is simply the number itself.

5.7 A deposit of £750 increases by 9%. Calculate the resulting deposit.

Solution We calculate 9% of 750

$$9\% \text{ of } 750 = \frac{9}{100} \times 750 = 67.5$$

The deposit has increased by £67.5. The resulting deposit is $750 + 67.5 = £817.5$.

Alternatively we may perform the calculation as follows. The original deposit represents 100%. The deposit increases by 9% to 109% of the original. So the resulting deposit is 109% of £750.

$$109\% \text{ of } 750 = \frac{109}{100} \times 750 = 817.5$$

5.8 A television set is advertised at £315. The retailer offers a 10% discount. How much do you pay for the television?

Solution

$$10\% \text{ of } 315 = \frac{10}{100} \times 315 = 31.5$$

The discount is £31.5 and so the cost is $315 - 31.5 = £283.5$.

Alternatively we can note that since the discount is 10%, then the selling price is 90% of the advertised price.

$$90\% \text{ of } £315 = \frac{90}{100} \times 315 = £283.5 \text{ as before}$$

Using a calculator

Many calculators have the facility to calculate percentages of quantities directly, although care must be taken with the order in which you enter the numbers. For example, to find 27% of 90, you might type

[9] [0] [×] [2] [7] [%]

and obtain the answer 24.3. Check that you can use your calculator to find percentages, and ask your tutor for help if necessary.

Self-assessment question 5.1

1. Give one reason why it is sometimes useful to express fractions as percentages.

Exercise 5.1

1. Calculate 23% of 124.
2. Express the following as percentages:

   (a) $\frac{9}{11}$ (b) $\frac{15}{20}$ (c) $\frac{9}{10}$ (d) $\frac{45}{50}$ (e) $\frac{75}{90}$
3. Express $\frac{13}{12}$ as a percentage.
4. Calculate 217% of 500.
5. A worker earns £400 a week. She receives a 6% increase. Calculate her new weekly wage.
6. A debt of £1200 is decreased by 17%. Calculate the remaining debt.
7. Express the following percentages as decimals:

   (a) 50% (b) 36% (c) 75%
   (d) 100% (e) 12.5%
8. A compact disc player normally priced at £256 is reduced in a sale by 20%. Calculate the sale price.
9. A bank deposit earns 7.5% interest in one year. Calculate the interest earned on a deposit of £15 000.

Ratio

Ratios are simply an alternative way of expressing fractions. Consider the problem of dividing £200 between 2 people, Ann and Bill, in the ratio of 7 : 3. This means that Ann receives £7 for every £3 that Bill receives. So every £10 is divided as £7 to Ann and £3 to Bill. Thus Ann receives $\frac{7}{10}$ of the money. Now $\frac{7}{10}$ of £200 is $\frac{7}{10} \times 200 = 140$. So Ann receives £140 and Bill receives £60.

WORKED EXAMPLE

5.9 Divide 170 in the ratio 3 : 2.

Solution A ratio of 3 : 2 means that every 5 parts are split as 3 and 2. That is, the first number is $\frac{3}{5}$ of the total; the second number is $\frac{2}{5}$ of the total.

$$\frac{3}{5} \text{ of } 170 = \frac{3}{5} \times 170 = 102$$

$$\frac{2}{5} \text{ of } 170 = \frac{2}{5} \times 170 = 68$$

The number is divided into 102 and 68.

Note from Worked Example 5.9 that to split a number in a given ratio we first find the total number of parts. The total number of parts is found by adding the numbers in the ratio. For example, if the ratio is given as $m : n$, the total number of parts is $m + n$. Then these $m + n$ parts are split into two with the first number being $\frac{m}{m+n}$ of the total, and the second number being $\frac{n}{m+n}$ of the total. Compare this with Worked Example 5.9.

WORKED EXAMPLE

5.10 Divide 250 cm in the ratio 1 : 3 : 4.

Solution Every 8 cm is divided into 1 cm, 3 cm and 4 cm. Thus the first length is $\frac{1}{8}$ of the total, the second length is $\frac{3}{8}$ of the total, and the third length is $\frac{4}{8}$ of the total.

$$\frac{1}{8} \text{ of } 250 = \frac{1}{8} \times 250 = 31.25$$

$$\frac{3}{8} \text{ of } 250 = \frac{3}{8} \times 250 = 93.75$$

$$\frac{4}{8} \text{ of } 250 = \frac{4}{8} \times 250 = 125$$

The 250 cm length is divided into 31.25 cm, 93.75 cm and 125 cm.

Ratios can be written in different ways. The ratio 3 : 2 can also be written as 6 : 4. This is clear if we note that 6 : 4 is a total of 10 parts split as $\frac{6}{10}$ and $\frac{4}{10}$ of the total. Since $\frac{6}{10}$ is equivalent to $\frac{3}{5}$, and $\frac{4}{10}$ is equivalent to $\frac{2}{5}$, we see that 6 : 4 is equivalent to 3 : 2.

Generally, any ratio can be expressed as an equivalent ratio by multiplying or dividing each term in the ratio by the same number. So, for example,

5 : 3 is equivalent to 15:9

and

$\frac{3}{4} : 2$ is equivalent to 3 : 8

WORKED EXAMPLES

5.11 Divide a mass of 380 kg in the ratio $\frac{3}{4} : \frac{1}{5}$.

Solution It is simpler to work with whole numbers, so first of all we produce an equivalent ratio by multiplying each term, first by 4, and then by 5, to give

$$\frac{3}{4} : \frac{1}{5} = 3 : \frac{4}{5} = 15 : 4$$

Note that this is equivalent to multiplying through by the lowest common multiple of 4 and 5.

So dividing 380 kg in the ratio $\frac{3}{4} : \frac{1}{5}$ is equivalent to dividing it in the ratio 15 : 4.

Now the total number of parts is 19 and so we split the 380 kg mass as

$$\frac{15}{19} \times 380 = 300$$

and

$$\frac{4}{19} \times 380 = 80$$

The total mass is split into 300 kg and 80 kg.

5.12 Bell metal, which is a form of bronze, is used for casting bells. It is an alloy of copper and tin. To manufacture bell metal requires 17 parts of copper to every 3 parts of tin.

(a) Express this requirement as a ratio.

(b) Express the amount of tin required as a percentage of the total.

(c) If the total amount of tin in a particular casting is 150 kg, find the amount of copper.

Solution

(a) Copper and tin are needed in the ratio 17 : 3.

(b) $\frac{3}{20}$ of the alloy is tin. Since $\frac{3}{20} = 15\%$ we find that 15% of the alloy is tin.

(c) A mass of 150 kg of tin makes up 15% of the total. So 1% of the total would have a mass of 10 kg. Copper, which makes up 85%, will have a mass of 850 kg.

Self-assessment question 5.2

1. Dividing a number in the ratio 2 : 3 is the same as dividing it in the ratio 10 : 15. True or false?

Exercise 5.2

1. Divide 180 in the ratio 8 : 1 : 3.
2. Divide 930 cm in the ratio 1 : 1 : 3.
3. A 6 m length of wood is cut in the ratio 2 : 3 : 4. Calculate the length of each piece.
4. Divide 1200 in the ratio 1 : 2 : 3 : 4.
5. A sum of £2600 is divided between Alan, Bill and Claire in the ratio of $2\frac{3}{4} : 1\frac{1}{2} : 2\frac{1}{4}$. Calculate the amount that each receives.
6. A mass of 40 kg is divided into three portions in the ratio 3 : 4 : 8. Calculate the mass of each portion.
7. Express the following ratios in their simplest forms:
   (a) 12 : 24 (b) 3 : 6 (c) 3 : 6 : 12
   (d) $\frac{1}{3} : 7$
8. A box contains two sizes of nails. The ratio of long nails to short nails is 2 : 7. Calculate the number of each type if the total number of nails is 108.

Test and assignment exercises 5

1. Express as decimals

   (a) 8% (b) 18% (c) 65%

2. Express as percentages

   (a) $\frac{3}{8}$ (b) $\frac{79}{100}$ (c) $\frac{56}{118}$

3. Calculate 27.3% of 1496.
4. Calculate 125% of 125.
5. Calculate 85% of 0.25.
6. Divided 0.5 in the ratio 2:4:9.
7. A bill totals £234.5 to which is added tax at 17.5%. Calculate the amount of tax to be paid.
8. An inheritance is divided between 3 people in the ratio 4:7:2. If the least amount received is £2300 calculate how much the other two people received.
9. Divide 70 in the ratio of 0.5:1.3:2.1.
10. Divide 50% in the ratio 2:3.

Percentage, ratio and dilution in biology

Test your understanding

1. You have counted 252 rabbits. You know they have a sex ratio of 4:2 male to female. How many of the rabbits you counted can be assumed to be male?

2. 60% of fish caught in one catch were cod, 20% were plaice and the remainder were halibut. Express this information as a ratio.

3. Out of a total of 48 butterflies spotted, 5 were Speckled Wood. Express this as a percentage.

4. You have 40 patients. 15 are male, the rest are female. Express this as a ratio.

5. Out of 238 children 3 showed symptoms of severe asthma, 14 showed mild symptoms, and the others showed no asthmatic symptoms. Express this data as percentages.

6. In a field of wheat measuring 2.6 hectares, 0.4 hectares was found to be infected with a rust fungus. Express this as a percentage of the total area of wheat.

7. Out of a colony of 640 swans 310 are female. Express this as a decimal.

8. Of an area of tissue stained with a fluorescent dye, 62% of the tissue was found to fluoresce. Express this as a decimal.

9. In a survey on garden birds, 32 birds were counted. 25 % were robins, 18 % were blackbirds, the rest were wrens. Express this as a ratio.

10. In a solution there are 0.3 mols NaCl, 0.6 moles KOH and 1.2 moles NaOH. Express this information as a ratio.

11. What is the molar concentration of a 10% (w/v) solution of NaCl? Molar mass of NaCl is 58.44 g mol^{-1}.

12. What mass of NaCl is contained in 50 mls of a 5% (w/v) solution?

Dilution

Often in biology you need to dilute a solution. It may be a dilution series of a bacteria culture, for example, or a dilution of prey for a feeding experiment. It could be dilution of a nutrient or toxic chemical. It is important to know how to make a dilution series and to calculate the concentration of the substance you are diluting, or the concentrations of the dilutions you are creating.

Figure 5.1
Representation of a bacteria culture.

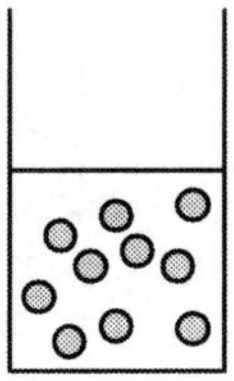

If this is 10 ml of culture where each circle represents 10^5 bacteria, and you are asked to make a 1/5 (a 1 in 5) dilution.....

You could take 1 ml of the culture..... and add it to 4 ml of medium

Figure 5.2
Diluting 1 ml of the bacteria culture

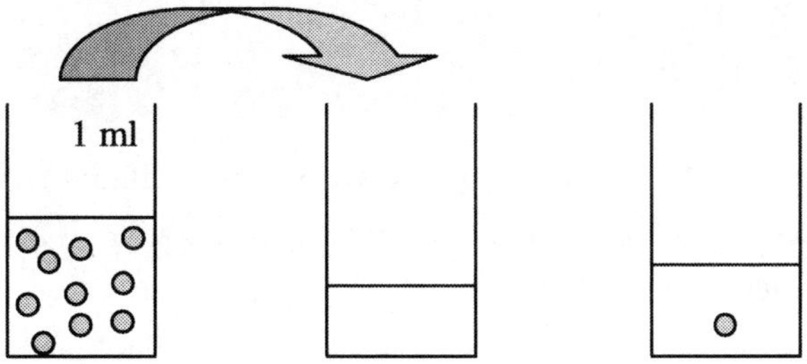

You will end up with 5 ml of culture, 1/5th of which was the original culture.

You could take 2 ml of the culture..... and add it to 8 ml of medium

Figure 5.3
Diluting 2 ml of the bacteria culture

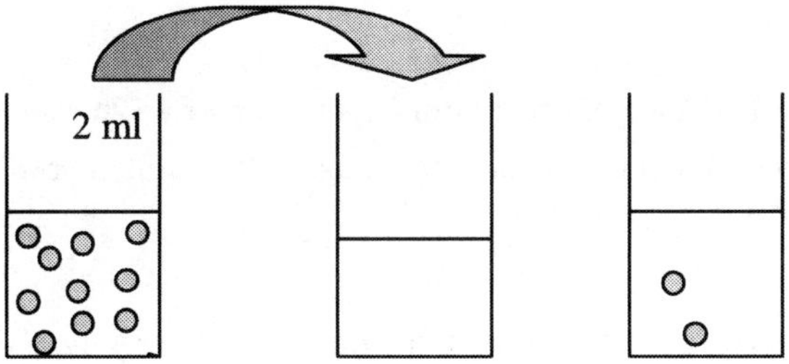

You will end up with 10 ml of culture, 1/5th of which was the original culture.
You will have twice as much as the previous dilution, but the concentration is the same for both.

The important values, when working with dilutions, are the properties of the original solution and the amount by which you are diluting that property.

A common exercise is to create a dilution series. This is often where a solution is continually diluted by a factor of 10 each time. Using a bacteria culture as an example, from a 10 ml culture, 1 ml is added to 9 ml of medium. This creates 10 ml of culture from which 1 ml is taken and added to another 9 ml of medium, and so on:

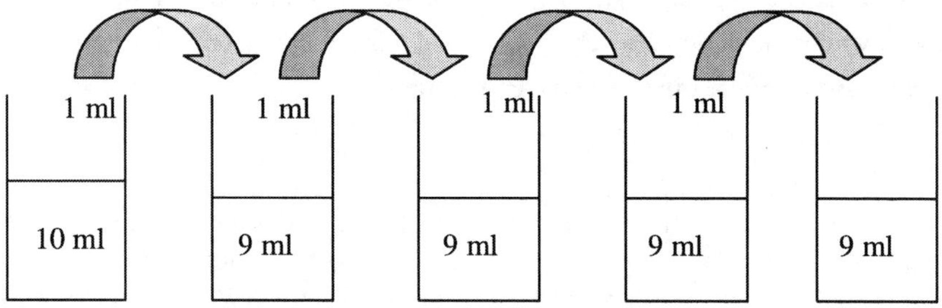

Figure 5.4 A dilution series

Resulting bacterial concentrations:

10^8 cells ml^{-1} 10^7 cells ml^{-1} 10^6 cells ml^{-1} 10^5 cells ml^{-1} 10^4 cells ml^{-1}

Each new bacterial solution is 1/10th the concentration of the previous one, because each is a 1/10 dilution.

Worked examples:

1) A protein solution of unknown concentration was diluted 1/5 (a five fold dilution). A protein assay was performed on the diluted protein sample and it was found to have a concentration of 2 mg ml^{-1}. What was the concentration of the original protein solution?

If the diluted sample was 1/5th of the concentration of the original solution (a 1/5 dilution), then you just multiply the concentration by five to find the original concentration:

Answer = 2 mg ml^{-1} x 5 = 10mg ml^{-1}

2) A solution of bromophenol blue had an absorbance reading at 600 nm of 1. This solution was diluted 1/10 (a ten fold dilution). What do you predict the absorbance to be of the diluted bromophenol blue solution?

If you dilute a solution 1/10, its properties such as absorption, will also decrease by 1/10th

So absorption of $1 \div 10 = 0.1$

3) Diluting prey for a feeding experiment

Imagine you had made a dilution series of 1/1, 1/2, 1/4, 1/6 prey cells ml^{-1}. If these prey concentrations gave you feeding rates of 36, 25, 9 and 6 cells $predator^{-1}$ ml^{-1} respectively, how would you represent this on a graph?

You would NOT put the dilution series on the *x*-axis. You would put the actual concentrations of prey on the axis. If the most concentrated prey sample was 6000 cells ml^{-1}, the dilutions would be 6000, 3000, 1500, 1000 cells ml^{-1} respectively. These concentrations of prey would then be used on a graph:

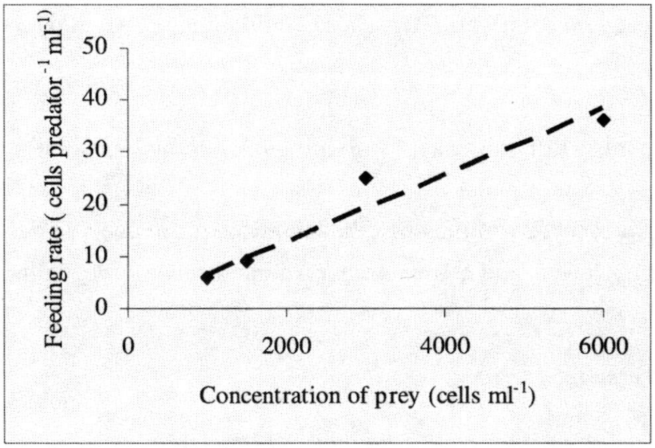

Figure 5.5 Feeding rate against prey concentration

How would you express these dilutions as percentages of the original prey concentration?

Dilutions are 1/1, 1/2, 1/4 and 1/6. This question simply expects you to convert these fractions into percentages:

1/1 = 100 %

1/2 = 50 %

1/4 = 25 %

1/6 = 16.7%

So, for example, the ¼ dilution gives you a prey concentration that is 25 % of the concentration of the original culture.

Scaling up and down

Often in biology you need to represent one measurement or size by another. For example, a cell is very small and if you make a drawing of it you represent its size using a scale bar. Or you might wish to represent the length of a gene in a diagram and give an idea of its relative size.

Relative scales can be represented as ratios. You could be told that a diagram is drawn at a scale of 10,000:1, where 1 μm, in reality, is represented on the page by 10,000 μm, or 10 mm. This means that 10 mm on the page is equivalent to 1 μm in real life. You could be asked to measure something on a diagram and say what its actual measurement is.

For example, if a picture of a cell was drawn to a scale of 10,000:1, then 10,000mm on the drawing would be 1 mm on the cell, or 1mm on the drawing would be 0.0001 mm on the cell. If the mitochondria measured 0.8 mm on the picture, what is its actual size?

Knowing that 1 mm is equivalent to 0.0001mm, you could multiply 0.8 mm by 0.0001, to give 0.00008 mm, or 80 nm.

Or you could calculate this a different way, by dividing 0.8 mm by 10,000 to give 0.00008 mm or 80 nm.

When working with microscopes you may be asked to measure an object under the microscope. To do this you have a scale bar in the eye piece, and a 1mm scale on a microscope slide, called a graticule.

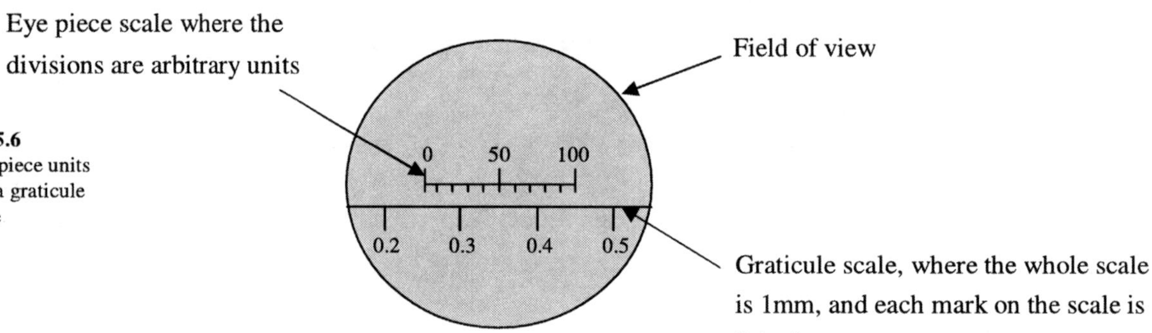

Fig 5.6
Eye piece units and a graticule scale

The process is the same as the scaling question above, and in the same way as above, there are two different ways to approach this. You could work out how many units in the eye-piece scale represent 1 mm on the slide, or graticule, or you could calculate what each unit on the eye piece represents.

On the eye piece scale, 50 units is equal to 0.1 mm, so 500 units would be equal to 1mm. So what ever you measure can be referred to as a proportion of 500 units, and so a proportion of 1mm.

In the diagram above, the total eye piece scale measures 0.2 mm. So each one unit on the eye piece will be equivalent to 0.002 mm, or 2 μm.

Taking another example, you might find that 100 units in the eye piece measures 0.1 mm. If you were to measure the length of a cell using the eye piece scale, and found it to be 45 units long, how long is this cell?

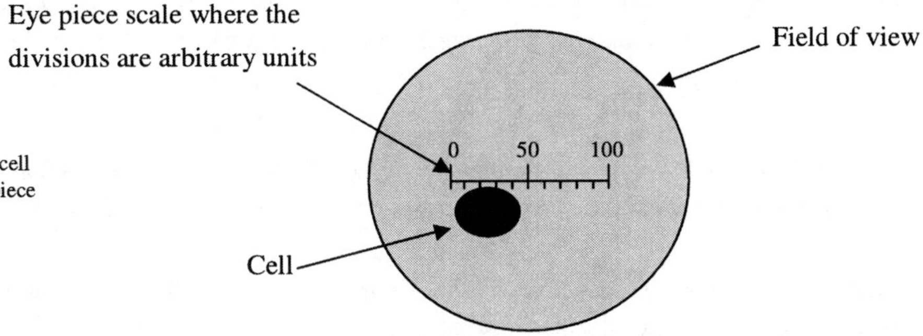

Fig 5.7
Measuring a cell against eye-piece units

First of all you have to work out the size of one eye piece unit, so divide 0.1 mm by 100 eye piece units. This means that each eye piece unit is 0.001 mm. This can be thought of as a scale of 1000:1, or where 1000 eye-piece units is equivalent to 1 mm.

So to find the length of the cell you would multiply 45 units by 0.001 mm, to give 0.045 mm or 45 μm.

How would you represent the size of this cell in a drawing?
For all drawings you should give some indication of size, either stating the actual size of the object being drawn, or adding a scale bar to your drawing.

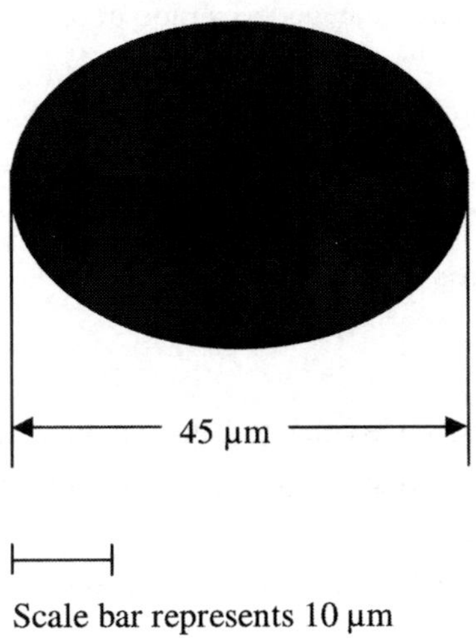

Fig 5.6
Representation of a diagram of a cell

This drawing shows a cell of 45 μm length in a drawing that has a length of 4.5 cm, or 45 mm.

You could say the diagram is drawn to a scale of 1000:1, meaning 1000 mm on the drawing is 1 mm of the actual cell.

The scale bar is 1 cm long and represents 10 μm.

Eye piece scales are necessary in microscopy because no microscope is the same as another, and calculating the size of an object that is under the microscope is specific not only to the microscope, but also to the objective you are using. So you need to calibrate the eye piece scale for each objective on each microscope you use.

Test your understanding

13. A protein solution of unknown concentration was diluted 1/8. A protein assay was performed on the diluted protein sample and it was found to have a concentration of 3 mg ml^{-1}. What was the concentration of the original protein solution?

14. A solution of bromophenol blue had an absorbance reading at 600 nm of 1.2. This solution was diluted 1/5. What do you predict to be the absorbance of the diluted bromophenol blue solution?

15. You are provided with a bacterial suspension with a concentration of 10^5 cells ml^{-1}. By how much would you dilute this suspension to achieve a concentration of 5×10^3 bacteria cells ml^{-1}?

16. You have made 1/2, 1/4 and 1/6 dilutions of a protist culture with 3×10^4 cells ml^{-1}. What is the concentration of protists in each of the dilutions?

17. A 10 mg ml^{-1} solution of DNA was diluted 1/5. What is the new concentration of the DNA in mg ml^{-1}?

18. You need to use 20 pg μl^{-1} of a DNA primer in a PCR reaction. You are supplied with the DNA primer at a concentration of 100 pg μl^{-1}. How much do you need to dilute the DNA primer by to get 20 pg μl^{-1}?

19. Calculate the concentration of bacteria resulting from a 1/6 dilution of a stock of 4×10^3 bacteria ml^{-1}. Express this as a percentage of the original concentration, and as a ratio of original:diluted concentrations.

20. A cell was represented by a diagram with a scale bar of 2 cm representing 2 μm. If the cell diagram had a diameter of 12 cm, how big is the cell?

21. A photograph had a scale of 10000:1. If a chloroplast measured 5 mm on this photograph, what would be the actual measurement of this organelle?

22. If the eye piece on a microscope had a scale of 35 units equal to 1 mm on the graticule slide, and the leg of an arthropod had a diameter of 12 eye piece units, what would the diameter of the arthropod's leg be?

23. A woodland 10 km across is represented on a photograph as being 10 cm across. What is the scale of this photograph?

24. In the question above (no. 23), if you measured the distance between two oak trees on the photograph as being 15 mm, what is the distance between them in the woodland?

Chapter 6
Algebra

03

Algebra

Objectives This chapter:

- explains what is meant by 'algebra'
- introduces important algebraic notations
- explains what is meant by a 'power' or 'index'
- illustrates how to evaluate an expression
- explains what is meant by a 'formula'

What is algebra?

In order to extend the techniques of arithmetic so that they can be more useful in applications we introduce letters or **symbols** to represent quantities of interest. For example, we may choose the capital letter I to stand for the *interest rate* in a business calculation, or the lower case letter t to stand for the *time* in a scientific calculation, and so on. The choice of which letter to use for which quantity is largely up to the user, although some conventions have been developed. Very often the letters x and y are used to stand for arbitrary quantities. **Algebra** is the body of mathematical knowledge that has been developed to manipulate symbols. Some symbols take fixed and unchanging values, and these are known as **constants**. For example, suppose we let the symbol b stand for the boiling point of water. This is fixed at 100 °C and so b is a constant. Some symbols represent quantities that can vary, and these are called **variables**. For example, the velocity of a car might be represented by the symbol v, and might vary from 0 to 100 kilometres per hour.

Algebraic notation

In algebraic work particular attention must be paid to the type of symbol used, so that, for example, the symbol T is quite different from the symbol t.

Table 6.1
The Greek alphabet

A	α	alpha	I	ι	iota	P	ρ	rho
B	β	beta	K	κ	kappa	Σ	σ	sigma
Γ	γ	gamma	Λ	λ	lambda	T	τ	tau
Δ	δ	delta	M	μ	mu	Y	υ	upsilon
E	ε	epsilon	N	ν	nu	Φ	ϕ	phi
Z	ζ	zeta	Ξ	ξ	xi	X	χ	chi
H	η	eta	O	o	omicron	Ψ	ψ	psi
Θ	θ	theta	Π	π	pi	Ω	ω	omega

Your scientific calculator is pre-programmed with the value of π. Check that you can use it.

Usually the symbols chosen are letters from the English alphabet although we frequently meet Greek letters. You may already be aware that the Greek letter 'pi', which has the symbol π, is used in the formula for the area of a circle, and is equal to the constant 3.14159.... In many calculations π can be approximated by $\frac{22}{7}$. For reference the full Greek alphabet is given in Table 6.1.

Another important feature is the position of a symbol in relation to other symbols. As we shall see in this chapter, the quantities xy, x^y, y^x and x_y all can mean quite different things. When a symbol is placed to the right and slightly higher than another symbol it is referred to as a **superscript**. So the quantity x^y contains the superscript y. Likewise, if a symbol is placed to the right and slightly lower than another symbol it is called a **subscript**. The quantity x_1 contains the subscript 1.

The arithmetic of symbols

Addition (+) If the letters x and y stand for two numbers, their **sum** is written as $x + y$. Note that $x + y$ is the same as $y + x$ just as $4 + 7$ is the same as $7 + 4$.

Subtraction (−) The quantity $x - y$ is called the **difference** of x and y, and means the number y subtracted from the number x. Note that $x - y$ is not the same as $y - x$, in the same way that $5 - 3$ is different from $3 - 5$.

Multiplication (×) Five times the number x is written $5 \times x$, although when multiplying the × sign is sometimes replaced with '.', or is even left out altogether. This means that $5 \times x$, $5.x$ and $5x$ all mean five times the number x. Similarly $x \times y$ can be written $x.y$ or simply xy. When multiplying, the order of the symbols is not important, so that xy is the same as yx just as 5×4 is the same as 4×5. The quantity xy is also known as the **product** of x and y.

Division (÷) $x \div y$ means the number x divided by the number y. This is also written x/y. Here the order is important and x/y is quite different from y/x. An expression involving one symbol divided by another is known as an **algebraic fraction**. The top line is called the **numerator** and the bottom line is called the **denominator**. The quantity x/y is known as the **quotient** of x and y.

A quantity made up of symbols together with $+$, $-$, $\times$ or $\div$ is called an **algebraic expression**. When evaluating an algebraic expression the BODMAS rule given in Chapter 1 applies. This rule reminds us of the correct order in which to evaluate an expression.

Self-assessment questions 6.1

1. Explain what you understand by the term 'algebra'.
2. If m and n are two numbers, explain what is meant by mn.
3. What is an algebraic fraction? Explain the meaning of the terms 'numerator' and 'denominator'.
4. What is the distinction between a superscript and a subscript?
5. What is the distinction between a variable and a constant?

Powers or indices

Frequently we shall need to multiply a number by itself several times, for example $3 \times 3 \times 3$, or $a \times a \times a \times a$.

To abbreviate such quantities a new notation is introduced. $a \times a \times a$ is written a^3, pronounced 'a cubed'. The superscript 3 is called a **power** or **index** and the letter a is called the **base**. Similarly $a \times a$ is written a^2, pronounced 'a squared' or 'a raised to the power 2'.

The power is the number of times that the base is multiplied by itself. Most calculators have a button marked x^y, which can be used to evaluate expressions such as 2^8, 3^{11} and so on. Check to see whether your calculator can do these by verifying that $2^8 = 256$ and $3^{11} = 177147$. Note that the plural of index is **indices**.

The calculator button x^y is used to find powers of numbers.

As a^2 means $a \times a$, and a^3 means $a \times a \times a$, then we interpret a^1 as simply a. That is, any number raised to the power 1 is itself.

Key point

Any number raised to the power 1 is itself, that is, $a^1 = a$.

WORKED EXAMPLES

6.1 In the expression 3^8 identify the index and the base.

Solution In the expression 3^8, the index is 8 and the base is 3.

6.2 Explain what is meant by y^5.

Solution y^5 means $y \times y \times y \times y \times y$.

6.3 Explain what is meant by x^2y^3.

Solution x^2 means $x \times x$; y^3 means $y \times y \times y$. Therefore x^2y^3 means $x \times x \times y \times y \times y$.

6.4 Evaluate 2^3 and 3^4.

Solution 2^3 means $2 \times 2 \times 2$, that is, 8. Similarly 3^4 means $3 \times 3 \times 3 \times 3$, that is, 81.

6.5 Explain what is meant by 7^1.

Solution Any number to the power 1 is itself, that is, 7^1 is simply 7.

6.6 Evaluate 10^2 and 10^3.

Solution 10^2 means 10×10 or 100. Similarly 10^3 means $10 \times 10 \times 10$ or 1000.

6.7 Use indices to write the expression $a \times a \times b \times b \times b$ more compactly.

Solution $a \times a$ can be written a^2; $b \times b \times b$ can be written b^3. Therefore $a \times a \times b \times b \times b$ can be written as $a^2 \times b^3$ or simply a^2b^3.

6.8 Write out fully z^3y^2.

Solution z^3y^2 means $z \times z \times z \times y \times y$. Note that we could also write this as $zzzyy$.

We now consider how to deal with expressions involving not only powers but other operations as well. Recall from the last section that the BODMAS rule tells us the order in which operations should be carried out, but the rule makes no reference to powers. In fact, powers should be given higher priority than any other operation and evaluated first. Consider the expression -4^2. Because the power must be evaluated first -4^2 is equal to -16. On the other hand $(-4)^2$ means $(-4) \times (-4)$ which is equal to $+16$.

WORKED EXAMPLES

6.9 Simplify (a) -5^2, (b) $(-5)^2$.

Solution

(a) The power is evaluated first. Noting that $5^2 = 25$, we see that $-5^2 = -25$.

(b) $(-5)^2$ means $(-5) \times (-5) = +25$.

Recall that when a negative number is multiplied by another negative number the result is positive.

Note how the brackets can significantly change the meaning of an expression.

6.10 Explain the meanings of $-x^2$ and $(-x)^2$. Are these different?

Solution In the expression $-x^2$ it is the quantity x that is squared, so that $-x^2 = -(x \times x)$. On the other hand $(-x)^2$ means $(-x) \times (-x)$, which equals $+x^2$. The two expressions are not the same.

Following the previous two examples we emphasise again the importance of the position of brackets in an expression.

Self-assessment questions 6.2

1. Explain the meaning of the terms 'power' and 'base'.
2. What is meant by an index?
3. Explain the distinction between $(xyz)^2$ and xyz^2.
4. Explain the distinction between $(-3)^4$ and -3^4.

Exercise 6.1

1. Evaluate the following without using a calculator: 2^4, $(\frac{1}{2})^2$, 1^8, 3^5 and 0^3.
2. Evaluate 10^4, 10^5 and 10^6 without using a calculator.
3. Use a calculator to evaluate 11^4, 16^8, 39^4 and 1.5^7.
4. Write out fully (a) a^4b^2c and (b) xy^2z^4.
5. Write the following expressions compactly using indices:
   (a) $xxxyyx$ (b) $xxyyzzz$
   (c) $xyzxyz$ (d) $abccba$
6. Using a calculator, evaluate
   (a) 7^4 (b) 7^5 (c) $7^4 \times 7^5$ (d) 7^9
   (e) 8^3 (f) 8^7 (g) $8^3 \times 8^7$ (h) 8^{10}
   Can you spot a rule for multiplying numbers with powers?
7. Without using a calculator, find $(-3)^3$, $(-2)^2$, $(-1)^7$ and $(-1)^4$.
8. Use a calculator to find $(-16.5)^3$, $(-18)^2$ and $(-0.5)^5$.
9. Without using a calculator find
   (a) $(-6)^2$ (b) $(-3)^2$ (c) $(-4)^3$
   (d) $(-2)^3$.
   Carefully compare your answers with the results of finding -6^2, -3^2, -4^3 and -2^3.

Substitution and formulae

Substitution means replacing letters by actual numerical values.

WORKED EXAMPLES

6.11 **Find the value of a^4 when $a = 3$.**

Solution a^4 means $a \times a \times a \times a$. When we **substitute** the number 3 in place of the letter a we find 3^4 or $3 \times 3 \times 3 \times 3$, that is, 81.

6.12 **Find the value of $a + 7b + 3c$ when $a = 1$, $b = 2$ and $c = 3$.**

Solution Letting $b = 2$ we note that $7b = 14$. Letting $c = 3$ we note that $3c = 9$. Therefore, with $a = 1$,

$$a + 7b + 3c = 1 + 14 + 9 = 24$$

6.13 **If $x = 4$, find the value of (a) $8x^3$ and (b) $(8x)^3$.**

Solution
(a) Substituting $x = 4$ into $8x^3$ we find $8 \times 4^3 = 8 \times 64 = 512$.
(b) Substituting $x = 4$ into $(8x)^3$ we obtain $(32)^3 = 32\ 768$. Note that the use of brackets makes a significant difference to the result.

6.14 **Evaluate mk, mn and nk when $m = 5$, $n = -4$ and $k = 3$.**

Solution $mk = 5 \times 3 = 15$. Similarly $mn = 5 \times (-4) = -20$ and $nk = (-4) \times 3 = -12$.

6.15 **Find the value of $-7x$ when (a) $x = 2$ and (b) $x = -2$.**

Solution
(a) Substituting $x = 2$ into $-7x$ we find -7×2, which equals -14.
(b) Substituting $x = -2$ into $-7x$ we find -7×-2, which equals 14.

6.16 **Find the value of x^2 when $x = -3$.**

Solution Because x^2 means $x \times x$, its value when $x = -3$ is -3×-3, that is, $+9$.

6.17 **Find the value of $-x^2$ when $x = -3$.**

Solution Recall that a power is evaluated first. So $-x^2$ means $-(x \times x)$. When $x = -3$, this evaluates to $-(-3 \times -3) = -9$.

6.18 **Find the value of $x^2 + 3x$ when (a) $x = 2$, (b) $x = -2$.**

Solution
(a) Letting $x = 2$ we find

$$x^2 + 3x = (2)^2 + 3(2) = 4 + 6 = 10$$

(b) Letting $x = -2$ we find

$$x^2 + 3x = (-2)^2 + 3(-2) = 4 - 6 = -2$$

6.19 **Find the value of $\frac{3x^2}{4} + 5x$ when $x = 2$.**

Solution Letting $x = 2$ we find

$$\frac{3x^2}{4} + 5x = \frac{3(2)^2}{4} + 5(2)$$

$$= \frac{12}{4} + 10$$

$$= 13$$

6.20 Find the value of $\frac{x^3}{4}$ when $x = 0.5$.

Solution When $x = 0.5$ we find

$$\frac{x^3}{4} = \frac{0.5^3}{4} = 0.03125$$

A **formula** is used to relate two or more quantities. You may already be familiar with the common formula used to find the area of a rectangle:

area = length × breadth

In symbols, writing A for area, l for length and b for breadth we have

$A = l \times b$ or simply $A = lb$

If we are now given particular numerical values for l and b we can use this formula to find A.

WORKED EXAMPLES

6.21 Use the formula $A = lb$ to find A when $l = 10$ and $b = 2.5$.

Solution Substituting the values $l = 10$ and $b = 2.5$ into the formula $A = lb$ we find $A = 10 \times 2.5 = 25$.

6.22 The formula $V = IR$ is used by electrical engineers. Find the value of V when $I = 12$ and $R = 7$.

Solution Substituting $I = 12$ and $R = 7$ in $V = IR$ we find $V = 12 \times 7 = 84$.

6.23 Use the formula $y = x^2 + 3x + 4$ to find y when $x = -2$.

Solution Substituting $x = -2$ into the formula gives

$$y = (-2)^2 + 3(-2) + 4 = 4 - 6 + 4 = 2$$

Self-assessment question 6.3

1. What is the distinction between an algebraic expression and a formula?

Exercise 6.2

1. Evaluate $3x^2y$ when $x = 2$ and $y = 5$.
2. Evaluate $8x + 17y - 2z$ when $x = 6$, $y = 1$ and $z = -2$.
3. The area A of a circle is found from the formula $A = \pi r^2$, where r is the length of the radius. Taking π to be 3.142 find the areas of the circles whose radii, in centimetres, are (a) $r = 10$, (b) $r = 3$, (c) $r = 0.2$.
4. Evaluate $3x^2$ and $(3x)^2$ when $x = 4$.
5. Evaluate $5x^2$ and $(5x)^2$ when $x = -2$.
6. If $y = 4.85$ find
   (a) $7y$ (b) y^2 (c) $5y + 2.5$
   (d) $y^3 - y$
7. If $a = 12.8$, $b = 3.6$ and $c = 9.1$ find
   (a) $a + b + c$ (b) ab (c) bc (d) abc
8. If $C = \frac{5}{9}(F - 32)$, find C when $F = 100$.
9. Evaluate (a) x^2, (b) $-x^2$ and (c) $(-x)^2$, when $x = 7$.
10. Evaluate the following when $x = -2$:
    (a) x^2 (b) $(-x)^2$ (c) $-x^2$
    (d) $3x^2$ (e) $-3x^2$ (f) $(-3x)^2$
11. Evaluate the following when $x = -3$:
    (a) $\frac{x^2}{3}$ (b) $(-x)^2$ (c) $-(\frac{x}{3})^2$
    (d) $4x^2$ (e) $-4x^2$ (f) $(-4x)^2$
12. Evaluate $x^2 - 7x + 2$ when $x = -9$.
13. Evaluate $2x^2 + 3x - 11$ when $x = -3$.
14. Evaluate $-x^2 + 3x - 5$ when $x = -1$.
15. Evaluate $-9x^2 + 2x$ when $x = 0$.
16. Evaluate $5x^2 + x + 1$ when (a) $x = 3$, (b) $x = -3$, (c) $x = 0$, (d) $x = -1$.
17. Evaluate $\frac{2x^2}{3} - \frac{x}{2}$ when
    (a) $x = 6$ (b) $x = -6$ (c) $x = 0$
    (d) $x = 1$
18. Evaluate $\frac{4x^2}{5} + 3$ when
    (a) $x = 0$ (b) $x = 1$ (c) $x = 5$
    (d) $x = -5$
19. Evaluate $\frac{x^3}{2}$ when
    (a) $x = -1$ (b) $x = 2$ (c) $x = 4$
20. Use the formula $y = \frac{x^3}{2} + 3x^2$ to find y when
    (a) $x = 0$ (b) $x = 2$ (c) $x = 3$
    (d) $x = -1$
21. If $g = 2t^2 - 1$, find g when
    (a) $t = 3$ (b) $t = 0.5$ (c) $t = -2$
22. In business calculations, the simple interest earned on an investment, I, is calculated from the formula $I = Prn$, where P is the amount invested, r is the interest rate and n is the number of time periods. Evaluate I when
    (a) $P = 15\,000$, $r = 0.08$ and $n = 5$
    (b) $P = 12\,500$, $r = 0.075$ and $n = 3$.
23. An investment earning 'compound interest' has a value, S, given by $S = P(1 + r)^n$, where P is the amount invested, r is the interest rate and n is the number of time periods. Calculate S when
    (a) $P = 8250$, $r = 0.05$ and $n = 15$
    (b) $P = 125\,000$, $r = 0.075$ and $n = 11$.

Test and assignment exercises 6

1. Using a calculator, evaluate 44^3, 0.44^2 and 32.5^3.
2. Write the following compactly using indices:

   (a) $xxxyyyyy$ (b) $\dfrac{xxx}{yyyy}$ (c) a^2baab
3. Evaluate the expression $4x^3yz^2$ when $x = 2$, $y = 5$ and $z = 3$.
4. The circumference C of a circle that has a radius of length r is given by the formula $C = 2\pi r$. Find the circumference of the circle with radius 0.5 cm. Take $\pi = 3.142$.
5. Find (a) $21^2 - 16^2$, (b) $(21 - 16)^2$. Comment upon the result.
6. If $x = 4$ and $y = -3$, evaluate

   (a) xy (b) $\dfrac{x}{y}$ (c) $\dfrac{x^2}{y^2}$ (d) $\left(\dfrac{x}{y}\right)^2$
7. Evaluate $2x(x + 4)$ when $x = 7$.
8. Evaluate $4x^2 + 7x$ when $x = 9$.
9. Evaluate $3x^2 - 7x + 12$ when $x = -2$.
10. Evaluate $-x^2 - 11x + 1$ when $x = -3$.
11. The formula $I = V/R$ is used by engineers. Find I when $V = 10$ and $R = 0.01$.
12. Given the formula $A = 1/x$, find A when (a) $x = 1$, (b) $x = 2$, (c) $x = 3$.
13. From the formula $y = 1/(x^2 + x)$ find y when (a) $x = 1$, (b) $x = -1$, (c) $x = 3$.
14. Find the value of $(-1)^n$ (a) when n is an even natural number and (b) when n is an odd natural number. (A natural number is a positive whole number.)
15. Find the value of $(-1)^{n+1}$ (a) when n is an even natural number and (b) when n is an odd natural number.

Chapter 7
Indices

Indices

Objectives: This chapter:

- states three laws used for manipulating indices
- shows how expressions involving indices can be simplified using the three laws
- explains the use of negative powers
- explains square roots, cube roots and fractional powers
- revises multiplication and division by powers of 10
- explains 'scientific notation' for representing very large and very small numbers

The laws of indices

Recall from Chapter 6 that an index is simply a power and that the plural of index is indices. Expressions involving indices can often be simplified if use is made of the **laws of indices**.

The first law

$$a^m \times a^n = a^{m+n}$$

In words, this states that if two numbers involving the same base but possibly different indices are to be multiplied together, their indices are added. Note that this law can be applied only if both bases are the same.

Key point

The first law: $a^m \times a^n = a^{m+n}$.

WORKED EXAMPLES

7.1 Use the first law of indices to simplify $a^4 \times a^3$.

Solution Using the first law we have $a^4 \times a^3 = a^{4+3} = a^7$. Note that the same result could be obtained by actually writing out all the terms:

$$a^4 \times a^3 = (a \times a \times a \times a) \times (a \times a \times a) = a^7$$

7.2 Use the first law of indices to simplify $3^4 \times 3^5$.

Solution From the first law $3^4 \times 3^5 = 3^{4+5} = 3^9$.

7.3 Simplify $a^4a^7b^2b^4$.

Solution $a^4a^7b^2b^4 = a^{4+7}b^{2+4} = a^{11}b^6$. Note that only those quantities with the same base can be combined using the first law.

The second law

$$\frac{a^m}{a^n} = a^{m-n}$$

In words, this states that if two numbers involving the same base but possibly different indices are to be divided, their indices are subtracted.

Key point

The second law: $\dfrac{a^m}{a^n} = a^{m-n}$.

WORKED EXAMPLES

7.4 Use the second law of indices to simplify $\frac{a^5}{a^3}$.

Solution The second law states that we subtract the indices, that is,

$$\frac{a^5}{a^3} = a^{5-3} = a^2$$

7.5 Use the second law of indices to simplify $\frac{3^7}{3^4}$.

Solution From the second law $\frac{3^7}{3^4} = 3^{7-4} = 3^3$.

7.6 Using the second law of indices, simplify $\frac{x^3}{x^3}$.

Solution Using the second law of indices we have $\frac{x^3}{x^3} = x^{3-3} = x^0$. However, note that any expression divided by itself equals 1, and so $\frac{x^3}{x^3}$ must equal 1. We can conclude from this that any number raised to the power 0 equals 1.

Key point

Any number raised to the power 0 equals 1, that is, $a^0 = 1$.

WORKED EXAMPLE

7.7 Evaluate (a) 14^0 (b) 0.5^0.

Solution (a) Any number to the power zero equals 1 and so $14^0 = 1$.
(b) Similarly, $0.5^0 = 1$.

The third law

$$(a^m)^n = a^{mn}$$

If a number is raised to a power, and the result is itself raised to a power, then the two powers are multiplied together.

Key point The third law: $(a^m)^n = a^{mn}$.

WORKED EXAMPLES

7.8 Simplify $(3^2)^4$.

Solution The third law states that the two powers are multiplied:

$$(3^2)^4 = 3^{2\times 4} = 3^8$$

7.9 Simplify $(x^4)^3$.

Solution Using the third law:

$$(x^4)^3 = x^{4\times 3} = x^{12}$$

7.10 Remove the brackets from the expression $(2a^2)^3$.

Solution $(2a^2)^3$ means $(2a^2) \times (2a^2) \times (2a^2)$. We can write this as

$$2 \times 2 \times 2 \times a^2 \times a^2 \times a^2$$

or simply $8a^6$. We could obtain the same result by noting that both terms in the bracket, that is, the 2 and the a^2, must be raised to the power 3, that is,

$$(2a^2)^3 = 2^3(a^2)^3 = 8a^6$$

The result of the previous example can be generalised to any term of the form $(a^m b^n)^k$. To simplify such an expression we make use of the formula $(a^m b^n)^k = a^{mk} b^{nk}$

Key point $(a^m b^n)^k = a^{mk} b^{nk}$

WORKED EXAMPLE

7.11 Remove the brackets from the expression $(x^2y^3)^4$.

Solution Using the previous result we find

$$(x^2y^3)^4 = x^8y^{12}$$

We often need to use several laws of indices in one example.

WORKED EXAMPLES

7.12 Simplify $\frac{(x^3)^4}{x^2}$.

Solution $(x^3)^4 = x^{12}$ using the third law of indices

So

$\frac{(x^3)^4}{x^2} = \frac{x^{12}}{x^2} = x^{10}$ using the second law

7.13 Simplify $(t^4)^2(t^2)^3$

Solution $(t^4)^2 = t^8$, $(t^2)^3 = t^6$ using the third law

So

$(t^4)^2(t^2)^3 = t^8t^6 = t^{14}$ using the first law.

Self-assessment questions 7.1

1. State the three laws of indices.
2. Explain what is meant by a^0.
3. Explain what is meant by x^1.

Exercise 7.1

1. Simplify
   (a) $5^7 \times 5^{13}$ (b) $9^8 \times 9^5$
   (c) $11^2 \times 11^3 \times 11^4$
2. Simplify
   (a) $\frac{15^3}{15^2}$ (b) $\frac{4^{18}}{4^9}$ (c) $\frac{5^{20}}{5^{19}}$
3. Simplify
   (a) a^7a^3 (b) a^4a^5 (c) $b^{11}b^{10}b$
4. Simplify
   (a) $x^7 \times x^8$ (b) $y^4 \times y^8 \times y^9$
5. Explain why the laws of indices cannot be used to simplify $19^8 \times 17^8$.

6. Simplify
(a) $(7^3)^2$ (b) $(4^2)^8$ (c) $(7^9)^2$

7. Simplify $\dfrac{1}{(5^3)^8}$

8. Simplify
(a) $(x^2y^3)(x^3y^2)$ (b) $(a^2bc^2)(b^2ca)$

9. Remove the brackets from
(a) $(x^2y^4)^5$ (b) $(9x^3)^2$ (c) $(-3x)^3$
(d) $(-x^2y^3)^4$

10. Simplify
(a) $\frac{(z^2)^3}{z^3}$ (b) $\frac{(y^3)^2}{(y^2)^2}$ (c) $\frac{(x^3)^2}{(x^2)^3}$

Negative powers

Sometimes a number is raised to a negative power. This is interpreted as follows:

$$a^{-m} = \frac{1}{a^m}$$

This can also be rearranged and expressed in the form

$$a^m = \frac{1}{a^{-m}}$$

Key point

$$a^{-m} = \frac{1}{a^m}, \qquad a^m = \frac{1}{a^{-m}}$$

For example,

$$3^{-2} \text{ means } \frac{1}{3^2}\text{, that is, } \frac{1}{9}$$

Similarly,

$$\text{the number } \frac{1}{5^{-2}} \text{ can be written } 5^2\text{, or simply } 25$$

To see the justification for this, note that because any number raised to the power 0 equals 1 we can write

$$\frac{1}{a^m} = \frac{a^0}{a^m}$$

Using the second law of indices to simplify the right-hand side we obtain $\frac{a^0}{a^m} = a^{0-m} = a^{-m}$ so that $\frac{1}{a^m}$ is the same as a^{-m}.

WORKED EXAMPLES

7.14 Evaluate

(a) 2^{-5} (b) $\frac{1}{3^{-4}}$

Solution (a) $2^{-5} = \frac{1}{2^5} = \frac{1}{32}$ (b) $\frac{1}{3^{-4}} = 3^4$ or simply 81

7.15 Evaluate (a) 10^{-1} (b) 10^{-2}

Solution
(a) 10^{-1} means $\frac{1}{10^1}$, or simply $\frac{1}{10}$. It is important to recognise that 10^{-1} is therefore the same as 0.1.

(b) 10^{-2} means $\frac{1}{10^2}$ or $\frac{1}{100}$. So 10^{-2} is therefore the same as 0.01.

7.16 Rewrite each of the following expressions using only positive powers:

(a) 7^{-3} (b) x^{-5}

Solution
(a) 7^{-3} means the same as $\frac{1}{7^3}$. The expression has now been written using a positive power.

(b) $x^{-5} = \frac{1}{x^5}$.

7.17 Rewrite each of the following expressions using only positive powers:

(a) $\frac{1}{x^{-9}}$ (b) $\frac{1}{a^{-4}}$

Solution (a) $\frac{1}{x^{-9}} = x^9$ (b) $\frac{1}{a^{-4}} = a^4$

7.18 Rewrite each of the following using only negative powers:

(a) 6^8 (b) x^5 (c) z^a

Solution (a) $6^8 = \frac{1}{6^{-8}}$ (b) $x^5 = \frac{1}{x^{-5}}$ (c) $z^a = \frac{1}{z^{-a}}$

7.19 Simplify

(a) $x^{-2}x^7$ (b) $\frac{x^{-3}}{x^{-5}}$

Solution
(a) To simplify $x^{-2}x^7$ we can use the first law of indices to write it as $x^{-2+7} = x^5$.

(b) To simplify $\frac{x^{-3}}{x^{-5}}$ we can use the second law of indices to write it as $x^{-3-(-5)} = x^{-3+5} = x^2$.

7.20 Simplify (a) $(x^{-3})^5$ (b) $\frac{1}{(x^{-2})^2}$

Solution
(a) To simplify $(x^{-3})^5$ we can use the third law of indices and write it as $x^{-3\times 5} = x^{-15}$. The answer could also be written as $\frac{1}{x^{15}}$.

(b) Note that $(x^{-2})^2 = x^{-4}$ using the third law. So $\frac{1}{(x^{-2})^2} = \frac{1}{x^{-4}}$. This could also be written as x^4.

Self-assessment question 7.2

1. Explain how the negative power in a^{-m} is interpreted.

Exercise 7.2

1. Without using a calculator express each of the following as a proper fraction:
(a) 2^{-2} (b) 2^{-3} (c) 3^{-2} (d) 3^{-3}
(e) 5^{-2} (f) 4^{-2} (g) 9^{-1} (h) 11^{-2}
(i) 7^{-1}

2. Express each of the following as decimal fractions:
(a) 10^{-1} (b) 10^{-2} (c) 10^{-6} (d) $\frac{1}{10^2}$
(e) $\frac{1}{10^3}$ (f) $\frac{1}{10^4}$

3. Write each of the following using only a positive power:
(a) x^{-4} (b) $\frac{1}{x^{-5}}$ (c) x^{-7} (d) y^{-2}
(e) $\frac{1}{y^{-1}}$ (f) y^{-1} (g) y^{-2} (h) z^{-1}
(i) $\frac{1}{z^{-1}}$

4. Simplify the following using the laws of indices and write your results using only positive powers:
(a) $x^{-2}x^{-1}$ (b) $x^{-3}x^{-2}$ (c) x^3x^{-4}
(d) $x^{-4}x^9$ (e) $\frac{x^{-2}}{x^{11}}$ (f) $(x^{-4})^2$
(g) $(x^{-3})^3$ (h) $(x^2)^{-2}$

5. Simplify
(a) $a^{13}a^{-2}$ (b) $x^{-9}x^{-7}$ (c) $x^{-21}x^2x$
(d) $(4^{-3})^2$

6. Evaluate
(a) 10^{-3} (b) 10^{-4} (c) 10^{-5}

7. Evaluate $4^{-8}/4^{-6}$ and $3^{-5}/3^{-8}$ without using a calculator.

Square roots, cube roots and fractional powers

Square roots

Consider the relationship between the numbers 5 and 25. We know that $5^2 = 25$ and so 25 is the square of 5. Equivalently we say that 5 is a **square root** of 25. The symbol $\sqrt[2]{\ }$, or simply $\sqrt{\ }$, is used to denote a square root and we write

$$5 = \sqrt{25}$$

We can picture this as follows:

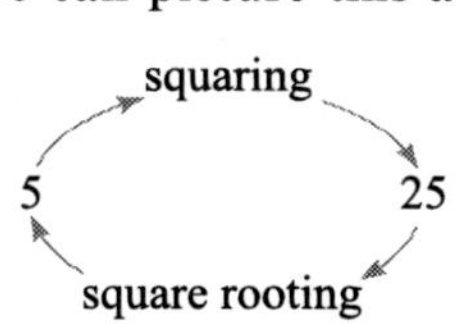

From this we see that taking the square root can be thought of as reversing the process of squaring.

We also note that

$$(-5) \times (-5) = (-5)^2$$
$$= 25$$

and so -5 is also a square root of 25. Hence we can write

$$-5 = \sqrt{25}$$

We can write both results together by using the 'plus or minus' sign $\pm$. We write

$$\sqrt{25} = \pm 5$$

In general, a **square root** of a number is a number that when squared gives the original number. Note that there are two square roots of any positive number but negative numbers possess no square roots.

Most calculators enable you to find square roots although only the positive value is normally given. Look for a $\sqrt{}$ button on your calculator.

WORKED EXAMPLE

7.21

(a) Use your calculator to find $\sqrt{79}$ correct to four decimal places.

(b) Check your answers are correct by squaring them.

Solution

(a) Using the $\sqrt{}$ button on the calculator you should verify that

$$\sqrt{79} = 8.8882 \text{ (to four decimal places)}$$

The second square root is -8.8882. Thus we can write

$$\sqrt{79} = \pm 8.8882.$$

(b) Squaring either of the numbers ± 8.8882 we recover the original number, 79.

Cube roots

The **cube root** of a number is a number that when cubed gives the original number. The symbol for a cube root is $\sqrt[3]{}$. So, for example, since $2^3 = 8$ we can write $\sqrt[3]{8} = 2$.

We can picture this as follows:

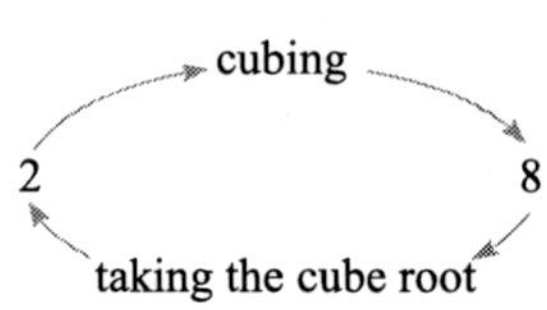

We can think of taking the cube root as reversing the process of cubing. As another example we note that $(-2)^3 = -8$ and hence $\sqrt[3]{-8} = -2$. All numbers, both positive and negative, possess a single cube root.

Your calculator may enable you to find a cube root. Look for a button marked $\sqrt[3]{\ }$. If so, check that you can use it correctly by verifying that

$$\sqrt[3]{46} = 3.5830$$

Fourth, fifth and other roots are defined in a similar way. For example, since

$$8^5 = 32\ 768$$

we can write

$$\sqrt[5]{32\ 768} = 8$$

Fractional powers

Sometimes fractional powers are used. The following example helps us to interpret a fractional power.

WORKED EXAMPLE

7.22 Simplify (a) $x^{\frac{1}{2}}x^{\frac{1}{2}}$ (b) $x^{\frac{1}{3}}x^{\frac{1}{3}}x^{\frac{1}{3}}$

Use your results to interpret the fractional powers $\frac{1}{2}$ and $\frac{1}{3}$.

Solution (a) Using the first law we can write

$$x^{\frac{1}{2}}x^{\frac{1}{2}} = x^{\frac{1}{2}+\frac{1}{2}} = x^1 = x$$

(b) Similarly,

$$x^{\frac{1}{3}}x^{\frac{1}{3}}x^{\frac{1}{3}} = x^{\frac{1}{3}+\frac{1}{3}+\frac{1}{3}} = x^1 = x$$

From (a) we see that

$$(x^{\frac{1}{2}})^2 = x$$

So when $x^{\frac{1}{2}}$ is squared, the result is x. Thus $x^{\frac{1}{2}}$ is simply the square root of x, that is,

$$x^{\frac{1}{2}} = \sqrt{x}$$

Similarly, from (b)

$$(x^{\frac{1}{3}})^3 = x$$

and so $x^{\frac{1}{3}}$ is the cube root of x, that is,

$$x^{\frac{1}{3}} = \sqrt[3]{x}$$

Key point $x^{\frac{1}{2}} = \sqrt{x}, \quad x^{\frac{1}{3}} = \sqrt[3]{x}$

More generally we have the following result:

Key point $x^{\frac{1}{n}} = \sqrt[n]{x}$

Your scientific calculator will probably be able to find fractional powers. The button may be marked $x^{1/y}$ or $\sqrt[y]{x}$. Check that you can use it correctly by working through the following examples.

WORKED EXAMPLES

7.23 Evaluate to three decimal places, using a calculator:

(a) $3^{\frac{1}{4}}$ (b) $15^{1/5}$

Solution Use your calculator to obtain the following solutions:

(a) 1.316 (b) 1.719

Note in part (a) that although the calculator gives just a single fourth root, there is another, −1.316.

7.24 Evaluate $(81)^{1/2}$.

Solution $(81)^{1/2} = \sqrt{81} = \pm 9$.

7.25 Explain what is meant by the number $27^{1/3}$.

Solution $27^{1/3}$ can be written $\sqrt[3]{27}$, that is, the cube root of 27. The cube root of 27 is 3, since $3 \times 3 \times 3 = 27$, and so $27^{1/3} = 3$. Note also that since $27 = 3^3$ we can write

$$(27)^{1/3} = (3^3)^{1/3} = 3^{(3 \times 1/3)} \quad \text{using the third law}$$
$$= 3^1 \quad = 3$$

The following worked example shows how we deal with negative fractional powers.

7.26 Explain what is meant by the number $(81)^{-1/2}$.

Solution Recall from our work on negative powers that $a^{-m} = 1/a^m$. Therefore we can write $(81)^{-1/2}$ as $1/(81)^{1/2}$. Now $81^{1/2} = \sqrt{81} = \pm 9$ and so

$$(81)^{-1/2} = \frac{1}{\pm 9} = \pm\frac{1}{9}$$

7.27 Write each of the following using a single index:

(a) $(5^2)^{\frac{1}{3}}$ (b) $(5^{-2})^{\frac{1}{3}}$

Solution (a) Using the third law of indices we find

$$(5^2)^{\frac{1}{3}} = 5^{2\times\frac{1}{3}} = 5^{\frac{2}{3}}$$

Note that $(5^2)^{\frac{1}{3}}$ is the cube root of 5^2, that is, $\sqrt[3]{25}$ or 2.9240.

(b) Using the third law of indices we find

$$(5^{-2})^{\frac{1}{3}} = 5^{-2\times\frac{1}{3}} = 5^{-\frac{2}{3}}$$

Note that there is a variety of equivalent ways in which this can be expressed, for example, $\sqrt[3]{\frac{1}{5^2}}$ or $\sqrt[3]{\frac{1}{25}}$, or as $\frac{1}{5^{2/3}}$.

7.28 Write each of the following using a single index:

(a) $\sqrt{x^3}$ (b) $(\sqrt{x})^3$

Solution (a) Because the square root of a number can be expressed as that number raised to the power $\frac{1}{2}$ we can write

$$\begin{aligned}\sqrt{x^3} &= (x^3)^{\frac{1}{2}}\\ &= x^{3\times\frac{1}{2}} \qquad \text{using the third law}\\ &= x^{\frac{3}{2}}\end{aligned}$$

(b)
$$\begin{aligned}(\sqrt{x})^3 &= (x^{\frac{1}{2}})^3\\ &= x^{\frac{3}{2}} \qquad \text{using the third law}\end{aligned}$$

Note from this example that $\sqrt{x^3} = (\sqrt{x})^3$.

Note that by generalising the results of the two previous worked examples we have the following:

Key point $a^{\frac{m}{n}} = \sqrt[n]{a^m} = = (\sqrt[n]{a})^m$

Self-assessment questions 7.3

1. Explain the meaning of the fractional powers $x^{1/2}$ and $x^{1/3}$.
2. What are the square roots of 100? Explain why the number -100 does not have any square roots.

Exercise 7.3

1. Evaluate
   (a) $64^{1/3}$ (b) $144^{1/2}$ (c) $16^{-1/4}$
   (d) $25^{-1/2}$ (e) $\dfrac{1}{32^{-1/5}}$
2. Simplify and then evaluate
   (a) $(3^{-1/2})^4$ (b) $(8^{1/3})^{-1}$
3. Write each of the following using a single index:
   (a) $\sqrt{8}$ (b) $\sqrt[3]{12}$ (c) $\sqrt[4]{16}$ (d) $\sqrt{13^3}$
   (e) $\sqrt[3]{4^7}$
4. Write each of the following using a single index:
   (a) $\sqrt{x}$ (b) $\sqrt[3]{y}$ (c) $\sqrt[2]{x^5}$ (d) $\sqrt[3]{5^7}$

Multiplication and division by powers of 10

To multiply and divide decimal fractions by powers of 10 is particularly simple. For example, to multiply 256.875 by 10 the decimal point is moved one place to the right, that is,

$$256.875 \times 10 = 2568.75$$

To multiply by 100 the decimal point is moved two places to the right. So

$$256.875 \times 100 = 25687.5$$

To divide a number by 10, the decimal point is moved one place to the left. This is equivalent to multiplying by 10^{-1}. To divide by 100, the decimal point is moved two places to the left. This is equivalent to multiplying by 10^{-2}.

In general, to multiply a number by 10^n, the decimal point is moved n places to the right if n is a positive integer, and n places to the left if n is a negative integer. If necessary, additional zeros are inserted to make up the required number of digits. Consider the following example.

WORKED EXAMPLE

7.29 Without the use of a calculator, write down

(a) 75.45×10^3 (b) 0.056×10^{-2} (c) 96.3×10^{-3} (d) 0.00743×10^5

Solution

(a) The decimal point is moved three places to the right: $75.45 \times 10^3 = 75450$. It has been necessary to include an additional zero to make up the required number of digits.

(b) The decimal point is moved two places to the left: $0.056 \times 10^{-2} = 0.00056$.

(c) $96.3 \times 10^{-3} = 0.0963$.

(d) $0.00743 \times 10^5 = 743$.

Exercise 7.4

1. Without the use of a calculator write down:
   (a) 7.43×10^2 (b) 7.43×10^4 (c) 0.007×10^4 (d) 0.07×10^{-2}
2. Write each of the following as a multiple of 10^2:
   (a) 300 (b) 356 (c) 32 (d) 0.57

Scientific notation

It is often necessary to use very large numbers such as 65000000000 or very small numbers such as 0.000000001. **Scientific notation** can be used to express such numbers in a more concise form, which avoids writing very lengthy strings of numbers. Each number is written in the form

$$a \times 10^n$$

where a is usually a number between 1 and 10. We also make use of the fact that

$$10 = 10^1, \qquad 100 = 10^2, \qquad 1000 = 10^3 \text{ and so on}$$

and also that

$$10^{-1} = \frac{1}{10} = 0.1, \qquad 10^{-2} = \frac{1}{100} = 0.01 \text{ and so on}$$

Then, for example,

the number 4000 can be written $4 \times 1000 = 4 \times 10^3$

Similarly

the number 68 000 can be written $6.8 \times 10\,000 = 6.8 \times 10^4$

and

the number 0.09 can be written $9 \times 0.01 = 9 \times 10^{-2}$

Note that all three numbers have been written in the form $a \times 10^n$ where a lies between 1 and 10.

WORKED EXAMPLES

7.30 Express the following numbers in scientific notation:

(a) 54 (b) −276 (c) 0.3

Solution (a) 54 can be written as 5.4×10, so in scientific notation we have 5.4×10^1.

(b) Negative numbers cause no problem: $-276 = -2.76 \times 10^2$.

(c) We can write 0.3 as 3×0.1 or 3×10^{-1}.

7.31 Write out fully the following numbers:

(a) 2.7×10^{-1} (b) 9.6×10^5 (c) -8.2×10^2

Solution (a) $2.7 \times 10^{-1} = 0.27$.

(b) $9.6 \times 10^5 = 9.6 \times 100\,000 = 960\,000$.

(c) $-8.2 \times 10^2 = -8.2 \times 100 = -820$.

7.32 Simplify the expression $(3 \times 10^2) \times (5 \times 10^3)$.

Solution The order in which the numbers are written down does not matter, and so we can write

$$(3 \times 10^2) \times (5 \times 10^3) = 3 \times 5 \times 10^2 \times 10^3 = 15 \times 10^5$$

Noting that $15 = 1.5 \times 10$ we can express the final answer in scientific notation:

$$15 \times 10^5 = 1.5 \times 10 \times 10^5 = 1.5 \times 10^6$$

Hence

$$(3 \times 10^2) \times (5 \times 10^3) = 1.5 \times 10^6$$

Self-assessment question 7.4

1. What is the purpose of using scientific notation?

Exercise 7.5

1. Express each of the following numbers in scientific notation:
(a) 45 (b) 45 000 (c) −450 (d) 90 000 000 (e) 0.15 (f) 0.00036 (g) 3.5 (h) −13.2 (i) 1 000 000 (j) 0.0975 (k) 45.34

2. Write out fully the following numbers:
(a) 3.75×10^2 (b) 3.97×10^1 (c) 1.875×10^{-1} (d) -8.75×10^{-3}

3. Simplify each of the following expressions, writing your final answer in scientific notation:
(a) $(4 \times 10^3) \times (6 \times 10^4)$ (b) $(9.6 \times 10^4) \times (8.3 \times 10^3)$ (c) $(1.2 \times 10^{-3}) \times (8.7 \times 10^{-2})$
(d) $\dfrac{9.37 \times 10^4}{6.14 \times 10^5}$ (e) $\dfrac{4.96 \times 10^{-2}}{9.37 \times 10^{-5}}$

Test and assignment exercises 7

1. Simplify

   (a) $\dfrac{z^5}{z^{-5}}$ (b) z^0 (c) $\dfrac{z^8 z^6}{z^{14}}$

2. Evaluate

   (a) $0.25^{1/2}$ (b) $(4096)^{1/3}$ (c) $(2601)^{1/2}$ (d) $16^{-1/2}$

3. Simplify $\dfrac{x^8 x^{-3}}{x^{-5} x^2}$

4. Find the value of $(1/7)^0$.

5. Remove the brackets from

   (a) $(abc^2)^2$ (b) $(xy^2z^3)^2$ (c) $(8x^2)^{-3}$

6. Express each of the following numbers in scientific notation:

   (a) 5792 (b) 98.4 (c) 0.001 (d) −66.667

Chapter 8
Simplifying Algebraic Expressions

Simplifying algebraic expressions

Objectives: This chapter:

- describes a number of ways in which complicated algebraic expressions can be simplified

Addition and subtraction of like terms

Like terms are multiples of the same quantity. For example, $3y$, $72y$ and $0.5y$ are all multiples of y and so are like terms. Similarly, $5x^2$, $-3x^2$ and $\frac{1}{2}x^2$ are all multiples of x^2 and so are like terms. xy, $17xy$ and $-91xy$ are all multiples of xy and are therefore like terms. Like terms can be collected together and added or subtracted in order to simplify them.

WORKED EXAMPLES

8.1 Simplify $3x + 7x - 2x$.

Solution All three terms are multiples of x and so are like terms. Therefore $3x + 7x - 2x = 8x$.

8.2 Simplify $3x + 2y$.

Solution $3x$ and $2y$ are not like terms. One is a multiple of x and the other is a multiple of y. The expression $3x+2y$ cannot be simplified.

8.3 Simplify $x + 7x + x^2$.

Solution The like terms are x and $7x$. These can be simplified to $8x$. Then $x + 7x + x^2 = 8x + x^2$. Note that $8x$ and x^2 are not like terms and so this expression cannot be simplified further.

8.4 Simplify $ab + a^2 - 7b^2 + 9ab + 8b^2$.

Solution

The terms ab and $9ab$ are like terms. Similarly the terms $-7b^2$ and $8b^2$ are like terms. These can be collected together and then added or subtracted as appropriate. Thus

$$\begin{aligned} ab + a^2 - 7b^2 + 9ab + 8b^2 &= ab + 9ab + a^2 - 7b^2 + 8b^2 \\ &= 10ab + a^2 + b^2 \end{aligned}$$

Exercise 8.1

1. Simplify, if possible,
   (a) $5p - 10p + 11q + 8q$ (b) $-7r - 13s + 2r + z$ (c) $181z + 13r - 2$
   (d) $x^2 + 3y^2 - 2y + 7x^2$ (e) $4x^2 - 3x + 2x + 9$
2. Simplify
   (a) $5y + 8p - 17y + 9q$ (b) $7x^2 - 11x^3 + 14x^2 + y^3$ (c) $4xy + 3xy + y^2$
   (d) $xy + yx$ (e) $xy - yx$

Multiplying algebraic expressions and removing brackets

Recall that when multiplying two numbers together the order in which we write them is irrelevant. For example, both 5×4 and 4×5 equal 20.

When multiplying three or more numbers together the order in which we carry out the multiplication is also irrelevant. By this we mean, for example, that when asked to multiply $3 \times 4 \times 5$ we can think of this as either $(3 \times 4) \times 5$ or as $3 \times (4 \times 5)$. Check for yourself that the result is the same, 60, either way.

It is also important to appreciate that $3 \times 4 \times 5$ could have been written as $(3)(4)(5)$.

It is essential that you grasp these simple facts about numbers in order to understand the algebra that follows. This is because identical rules are applied. Rules for determining the sign of the answer when multiplying positive and negative algebraic expressions are also the same as those used for multiplying numbers.

Key point

When multiplying

positive × positive = positive
positive × negative = negative
negative × positive = negative
negative × negative = positive

We introduce the processes involved in removing brackets using some simple examples.

WORKED EXAMPLES

8.5 Simplify $3(4x)$.

Solution Just as with numbers $3(4x)$ could be written as $3 \times (4 \times x)$, and then as $(3 \times 4) \times x$, which evaluates to $12x$.
So $3(4x) = 12x$.

8.6 Simplify $5(3y)$.

Solution $5(3y) = 5 \times 3 \times y = 15y$.

8.7 Simplify $(5a)(3a)$.

Solution Here we can write $(5a)(3a) = (5 \times a) \times (3 \times a)$. Neither the order in which we carry out the multiplications nor the order in which we write down the terms matters, and so we can write this as

$$(5a)(3a) = (5 \times 3)(a \times a)$$

As we have shown, it is usual to write numbers at the beginning of an expression. This simplifies to $15 \times a^2$, that is $15a^2$. Hence

$$(5a)(3a) = 15a^2$$

8.8 Simplify $4x^2 \times 7x^5$.

Solution Recall that, when multiplying, the order in which we write down the terms does not matter. Therefore we can write

$$4x^2 \times 7x^5 = 4 \times 7 \times x^2 \times x^5$$

which equals $28x^{2+5} = 28x^7$.

8.9 Simplify $7(2b^2)$.

Solution $7(2b^2) = 7 \times (2 \times b^2) = (7 \times 2) \times b^2 = 14b^2$.

8.10 Simplify $(a) \times (-b)$

Solution Here we have the product of a positive and a negative quantity. The result will be negative. We write

$$(a) \times (-b) = -ab$$

8.11 Explain the distinction between ab^2 and $(ab)^2$.

Solution ab^2 means $a \times b \times b$ whereas $(ab)^2$ means $(ab) \times (ab)$ which equals $a \times b \times a \times b$. The latter could also be written as a^2b^2.

8.12 Simplify (a) $(6z)(8z)$ (b) $(6z) + (8z)$, noting the distinction between the two results.

Solution (a) $(6z)(8z) = 48z^2$.

(b) $(6z) + (8z)$ is the addition of like terms. This simplifies to $14z$.

8.13 Simplify (a) $(6x)(-2x)$ (b) $(-3y^2)(-2y)$.

Solution (a) $(6x)(-2x)$ means $(6x) \times (-2x)$, which equals $-12x^2$.

(b) $(-3y^2)(-2y) = (-3y^2) \times (-2y) = 6y^3$.

Self-assessment questions 8.1

1. Two negative expressions are multiplied together. State the sign of the resulting product.
2. Three negative expressions are multiplied together. State the sign of the resulting product.

Exercise 8.2

1. Simplify each of the following:
(a) (4)(3)(7) (b) (7)(4)(3) (c) (3)(4)(7)
2. Simplify
(a) $5 \times (4 \times 2)$ (b) $(5 \times 4) \times 2$
3. Simplify each of the following:
(a) $7(2z)$ (b) $15(2y)$ (c) $(2)(3)x$
(d) $9(3a)$ (e) $(11)(5a)$ (f) $2(3x)$
4. Simplify each of the following:
(a) $5(4x^2)$ (b) $3(2y^3)$ (c) $11(2u^2)$
(d) $(2 \times 4) \times u^2$ (e) $(13)(2z^2)$
5. Simplify
(a) $(7x)(3x)$ (b) $3a(7a)$ (c) $14a(a)$
6. Simplify
(a) $5y(3y)$ (b) $5y + 3y$
Explain why the two results are not the same.
7. Simplify the following:
(a) $(abc)(a^2bc)$ (b) $x^2y(xy)$
(c) $(xy^2)(xy^2)$
8. Explain the distinction, if any, between $(xy^2)(xy^2)$ and xy^2xy^2.
9. Explain the distinction, if any, between $(xy^2)(xy^2)$ and $(xy^2) + (xy^2)$.
In both cases simplify the expressions.
10. Simplify
(a) $(3z)(-7z)$ (b) $3z - 7z$
11. Simplify
(a) $(-x)(3x)$ (b) $-x + 3x$
12. Simplify
(a) $(-2x)(-x)$ (b) $-2x - x$

Removing brackets from $a(b + c)$, $a(b - c)$ and $(a + b)(c + d)$

Recall from your study of arithmetic that the expression $(5 - 4) + 7$ is different from $5 - (4 + 7)$ because of the position of the brackets. In order to simplify an expression it is often necessary to remove brackets.

Removing brackets from expressions of the form $a(b + c)$ and $a(b - c)$

In an expression such as $a(b + c)$, it is intended that the a multiplies all the bracketed terms:

Key point

$a(b + c) = ab + ac$ Similarly: $a(b - c) = ab - ac$

WORKED EXAMPLES

8.14 Remove the brackets from

(a) $6(x + 5)$ (b) $8(2x - 4)$

Solution (a) In the expression $6(x + 5)$ it is intended that the 6 multiplies both terms in the brackets. Therefore

$$6(x + 5) = 6x + 30$$

(b) In the expression $8(2x - 4)$ the 8 multiplies both terms in the brackets so that

$$8(2x - 4) = 16x - 32$$

8.15 Remove the brackets from the expression $7(5x + 3y)$.

Solution The 7 multiplies both the terms in the bracket. Therefore

$$7(5x + 3y) = 7(5x) + 7(3y) = 35x + 21y$$

8.16 Remove the brackets from $-(x + y)$.

Solution The expression $-(x + y)$ actually means $-1(x + y)$. It is intended that the -1 multiplies both terms in the brackets, therefore

$$-(x + y) = -1(x + y) = (-1) \times x + (-1) \times y = -x - y$$

8.17 Remove the brackets from the expression

$$(x + y)z$$

Solution Note that the order in which we write down the terms to be multiplied does not matter, so that we can write $(x + y)z$ as $z(x + y)$. Then

$$z(x + y) = zx + zy$$

Alternatively note that $(x + y)z = xz + yz$, which is an equivalent form of the answer.

8.18 Remove the brackets from the expressions

(a) $5(x - 2y)$ (b) $(x + 3)(-1)$

Solution (a) $5(x - 2y) = 5x - 5(2y) = 5x - 10y$.

(b) $(x + 3)(-1) = (-1)(x + 3) = -1x - 3 = -x - 3$.

8.19 Simplify $x + 8(x - y)$.

Solution An expression such as this is simplified by first removing the brackets and then collecting together like terms. Removing the brackets we find

$$x + 8(x - y) = x + 8x - 8y$$

Collecting like terms we obtain $9x - 8y$.

8.20 Remove the brackets from

(a) $\frac{1}{2}(x + 2)$ (b) $\frac{1}{2}(x - 2)$ (c) $-\frac{1}{3}(a + b)$

Solution (a) In the expression $\frac{1}{2}(x + 2)$ it is intended that the $\frac{1}{2}$ multiplies both the terms in the brackets. So

$$\frac{1}{2}(x + 2) = \frac{1}{2}x + \frac{1}{2}(2) = \frac{1}{2}x + 1$$

(b) Similarly,

$$\frac{1}{2}(x - 2) = \frac{1}{2}x - \frac{1}{2}(2) = \frac{1}{2}x - 1$$

(c) In the expression $-\frac{1}{3}(a + b)$ the term $-\frac{1}{3}$ multiplies both terms in the brackets. So

$$-\frac{1}{3}(a + b) = -\frac{1}{3}a - \frac{1}{3}b$$

Removing brackets from expressions of the form $(a + b)(c + d)$

In the expression $(a + b)(c + d)$ it is intended that the quantity $(a + b)$ multiplies both the c and the d in the second bracket. Therefore

$$(a + b)(c + d) = (a + b)c \quad + \quad (a + b)d$$

Each of these two terms can be expanded further to give

$$(a + b)c = ac + bc \qquad \text{and} \qquad (a + b)d = ad + bd$$

Therefore

Key point

$$(a + b)(c + d) = ac + bc + ad + bd$$

WORKED EXAMPLES

8.21 Remove the brackets from $(3 + x)(2 + y)$.

Solution $(3+x)(2+y) = (3+x)(2) + (3+x)y$

$$= 6 + 2x + 3y + xy$$

8.22 Remove the brackets from $(x+6)(x-3)$.

Solution

$$(x+6)(x-3) = (x+6)x + (x+6)(-3)$$
$$= x^2 + 6x - 3x - 18$$
$$= x^2 + 3x - 18$$

8.23 Remove the brackets from

(a) $(1-x)(2-x)$ (b) $(-x-2)(2x-1)$

Solution

(a) $(1-x)(2-x) = (1-x)2 + (1-x)(-x)$
$$= 2 - 2x - x + x^2$$
$$= 2 - 3x + x^2$$

(b) $(-x-2)(2x-1) = (-x-2)(2x) + (-x-2)(-1)$
$$= -2x^2 - 4x + x + 2$$
$$= -2x^2 - 3x + 2$$

8.24 Remove the brackets from the expression $3(x+1)(x-1)$.

Solution First consider the expression $(x+1)(x-1)$.

$$(x+1)(x-1) = (x+1)x + (x+1)(-1)$$
$$= x^2 + x - x - 1$$
$$= x^2 - 1$$

Then $3(x+1)(x-1) = 3(x^2 - 1) = 3x^2 - 3$.

Exercise 8.3

1. Remove the brackets from
   (a) $4(x+1)$ (b) $-4(x+1)$
   (c) $4(x-1)$ (d) $-4(x-1)$
2. Remove the brackets from the following expressions:
   (a) $5(x-y)$ (b) $19(x+3y)$
   (c) $8(a+b)$ (d) $(5+x)y$
   (e) $12(x+4)$ (f) $17(x-9)$
   (g) $-(a-2b)$ (h) $\frac{1}{2}(2x+1)$
   (i) $-3m(-2+4m+3n)$
3. Remove the brackets and simplify the following:
   (a) $18-13(x+2)$ (b) $x(x+y)$
4. Remove the brackets and simplify the following expressions:
   (a) $(x+1)(x+6)$ (b) $(x+4)(x+5)$
   (c) $(x-2)(x+3)$ (d) $(x+6)(x-1)$
   (e) $(x+y)(m+n)$ (f) $(4+y)(3+x)$
   (g) $(5-x)(5+x)$
   (h) $(17x+2)(3x-5)$

5. Remove the brackets and simplify the following expressions:
(a) $(x+3)(x-7)$ (b) $(2x-1)(3x+7)$
(c) $(4x+1)(4x-1)$
(d) $(x+3)(x-3)$ (e) $(2-x)(3+2x)$

6. Remove the brackets and simplify the following expressions:
(a) $\frac{1}{2}(x+2y)+\frac{7}{2}(4x-y)$
(b) $\frac{3}{4}(x-1)+\frac{1}{4}(2x+8)$

7. Remove the brackets from
(a) $-(x-y)$ (b) $-(a+2b)$
(c) $-\frac{1}{2}(3p+q)$

8. Remove the brackets from $(x+1)(x+2)$. Use your result to remove the brackets from $(x+1)(x+2)(x+3)$.

Test and assignment exercises 8

1. Simplify
(a) $7x^2+4x^2+9x-8x$ (b) $y+7-18y+1$ (c) $a^2+b^2+a^3-3b^2$

2. Simplify
(a) $(3a^2b)\times(-a^3b^2c)$ (b) $\dfrac{x^3}{-x^2}$

3. Remove the brackets from
(a) $(a+3b)(7a-2b)$ (b) $x^2(x+2y)$ (c) $x(x+y)(x-y)$

4. Remove the brackets from
(a) $(7x+2)(3x-1)$ (b) $(1-x)(x+3)$ (c) $(5+x)x$ (d) $(8x+4)(7x-2)$

5. Remove the brackets and simplify
(a) $3x(x+2)-7x^2$ (b) $-(2a+3b)(a+b)$ (c) $4(x+7)+13(x-2)$
(d) $5(2a+5)-3(5a-2)$ (e) $\frac{1}{2}(a+4b)+\frac{3}{2}a$

Chapter 9
Algebraic Fractions

03

Algebraic fractions

Objectives: This chapter:

- explains how to simplify algebraic fractions by cancelling common factors
- explains how algebraic fractions can be multiplied and divided
- explains how algebraic fractions can be added and subtracted

Introduction

Just as one whole number divided by another is a numerical fraction, so one algebraic expression divided by another is called an **algebraic fraction**.

$$\frac{x}{y} \qquad \frac{x^2+y}{x} \qquad \frac{3x+2}{7}$$

are all examples of algebraic fractions. The top line is known as the **numerator** of the fraction, and the bottom line is the **denominator**.

Rules for determining the sign of the answer when dividing positive and negative algebraic expressions are the same as those used for dividing numbers.

Key point

When dividing

$$\frac{\text{positive}}{\text{positive}} = \text{positive} \qquad \frac{\text{negative}}{\text{positive}} = \text{negative}$$

$$\frac{\text{positive}}{\text{negative}} = \text{negative} \qquad \frac{\text{negative}}{\text{negative}} = \text{positive}$$

Using these rules we see that an algebraic expression can often be written in different but equivalent forms. For example note that

$$\frac{x}{-y} \text{ can be written as } -\frac{x}{y}$$

and that

$$\frac{-x}{y} \text{ can be written as } -\frac{x}{y}$$

and also that

$$\frac{-x}{-y} \text{ can be written as } \frac{x}{y}$$

Cancelling common factors

Cancellation of common factors was described in detail in §2.2.

Consider the numerical fraction $\frac{3}{12}$. To simplify this we factorise both the numerator and the denominator. Any factors which appear in both the numerator and the denominator are called **common factors**. These can be cancelled. For example,

$$\frac{3}{12} = \frac{1 \times 3}{4 \times 3} = \frac{1 \times \cancel{3}}{4 \times \cancel{3}} = \frac{1}{4}$$

The same process is applied when dealing with algebraic fractions.

WORKED EXAMPLES

9.1 For each pair of expressions, state which factors are common to both.

(a) $3xy$ and $6xz$ (b) xy and $5y^2$ (c) $3(x+2)$ and $(x+2)^2$
(d) $3(x-1)$ and $(x-1)(x+4)$

Solution

(a) The expression $6xz$ can be written $(3)(2)xz$. We see that factors common to both this and $3xy$ are 3 and x.

(b) The expression $5y^2$ can be written $5(y)(y)$. We see that the only factor common to both this and xy is y.

(c) $(x+2)^2$ can be written $(x+2)(x+2)$. Thus $(x+2)$ is a factor common to both $(x+2)^2$ and $3(x+2)$.

(d) $3(x-1)$ and $(x-1)(x+4)$ have a common factor of $(x-1)$.

9.2 Simplify

$$\frac{18x^2}{6x}$$

Solution First note that 18 can be factorised as 6×3. So there are factors of 6 and x in both the numerator and the denominator. Then common factors can be cancelled. That is,

$$\frac{18x^2}{6x} = \frac{(6)(3)x^2}{6x} = \frac{3x}{1} = 3x$$

Key point

When simplifying an algebraic fraction only factors common to both the numerator and denominator can be cancelled.

A fraction is expressed in its simplest form by factorising the numerator and denominator and cancelling any common factors.

WORKED EXAMPLES

9.3 Simplify

$$\frac{5}{25 + 15x}$$

Solution First of all note that the denominator can be factorised as $5(5 + 3x)$. There is therefore a factor of 5 in both the numerator and denominator. So 5 is a common factor. This can be cancelled. That is,

$$\frac{5}{25 + 15x} = \frac{1 \times 5}{5(5 + 3x)} = \frac{1 \times \cancel{5}}{\cancel{5}(5 + 3x)} = \frac{1}{5 + 3x}$$

It is very important to note that the number 5 that has been cancelled is a common factor. It is incorrect to try to cancel terms that are not common factors.

9.4 Simplify

$$\frac{5x}{25x + 10y}$$

Solution Factorising the denominator we can write

$$\frac{5x}{25x + 10y} = \frac{5x}{5(5x + 2y)}$$

We see that there is a common factor of 5 in both numerator and denominator that can be cancelled. Thus

$$\frac{5x}{25x+10y} = \frac{\not{5}x}{\not{5}(5x+2y)} = \frac{x}{5x+2y}$$

Note that no further cancellation is possible. x is not a common factor because it is not a factor of the denominator.

9.5 Simplify

$$\frac{4x}{3x^2+x}$$

Solution Note that the denominator factorises to $x(3x+1)$. Once both numerator and denominator have been factorised, any common factors are cancelled. So

$$\frac{4x}{3x^2+x} = \frac{4x}{x(3x+1)} = \frac{4\not{x}}{\not{x}(3x+1)} = \frac{4}{3x+1}$$

Note that the factor x is common to both numerator and denominator and so has been cancelled.

9.6 Simplify

$$\frac{x}{x^2+2x}$$

Solution Note that the denominator factorises to $x(x+2)$. Also note that the numerator can be written as $1 \times x$. So

$$\frac{x}{x^2+2x} = \frac{1 \times x}{x(x+2)} = \frac{1 \times \not{x}}{\not{x}(x+2)} = \frac{1}{x+2}$$

9.7 Simplify

(a) $\dfrac{2(x-1)}{(x+3)(x-1)}$ (b) $\dfrac{x-4}{(x-4)^2}$

Solution (a) There is a factor of $(x-1)$ common to both the numerator and denominator. This is cancelled to give

$$\frac{2(x-1)}{(x+3)(x-1)} = \frac{2}{x+3}$$

(b) There is a factor of $x-4$ in both numerator and denominator. This is cancelled as follows:

$$\frac{x-4}{(x-4)^2} = \frac{1(x-4)}{(x-4)(x-4)} = \frac{1}{x-4}$$

9.8 Simplify

$$\frac{x+2}{x^2+3x+2}$$

Solution The denominator is factorised and then any common factors are cancelled:

$$\frac{x+2}{x^2+3x+2} = \frac{1(x+2)}{(x+2)(x+1)} = \frac{1}{x+1}$$

9.9 Simplify

(a) $\dfrac{3x+xy}{x^2+5x}$ (b) $\dfrac{x^2-1}{x^2+3x+2}$

Solution The numerator and denominator are both factorised and any common factors are cancelled.

(a) $\dfrac{3x+xy}{x^2+5x} = \dfrac{x(3+y)}{x^2+5x} = \dfrac{\not{x}(3+y)}{\not{x}(x+5)} = \dfrac{3+y}{5+x}$

(b) $\dfrac{x^2-1}{x^2+3x+2} = \dfrac{(x+1)(x-1)}{(x+1)(x+2)} = \dfrac{x-1}{x+2}$

Self-assessment questions 9.1

1. Explain why no cancellation is possible in the expression $\dfrac{3x}{3x+y}$.
2. Explain why no cancellation is possible in the expression $\dfrac{x+1}{x+3}$.
3. Explain why it is possible to perform a cancellation in the expression $\dfrac{x+1}{2x+2}$, and perform it.

Exercise 9.1

1. Simplify
   (a) $\dfrac{9x}{3y}$ (b) $\dfrac{9x}{x^2}$ (c) $\dfrac{9xy}{3x}$ (d) $\dfrac{9xy}{3y}$
   (e) $\dfrac{9xy}{xy}$ (f) $\dfrac{9xy}{3xy}$
2. Simplify
   (a) $\dfrac{15x}{3y}$ (b) $\dfrac{15x}{5y}$ (c) $\dfrac{15xy}{x}$ (d) $\dfrac{15xy}{xy}$
   (e) $\dfrac{x^5}{-x^3}$ (f) $\dfrac{-y^3}{y^7}$ (g) $\dfrac{-y}{-y^2}$ (h) $\dfrac{-y^{-3}}{-y^4}$
3. Simplify the following algebraic fractions:
   (a) $\dfrac{4}{12+8x}$ (b) $\dfrac{5+10x}{5}$ (c) $\dfrac{2}{4+14x}$
   (d) $\dfrac{2x}{4+14x}$ (e) $\dfrac{2x}{2+14x}$ (f) $\dfrac{7}{49x+7y}$
   (g) $\dfrac{7y}{49x+7y}$ (h) $\dfrac{7x}{49x+7y}$

4. Simplify
   (a) $\dfrac{15x+3}{3}$ (b) $\dfrac{15x+3}{3x+6y}$ (c) $\dfrac{12}{4x+8}$
   (d) $\dfrac{12x}{4xy+8x}$ (e) $\dfrac{13x}{x^2+5x}$
   (f) $\dfrac{17y}{9y^2+4y}$
5. Simplify the following:
   (a) $\dfrac{5}{15+10x}$ (b) $\dfrac{2x}{x^2+7x}$
   (c) $\dfrac{2x+8}{x^2+2x-8}$ (d) $\dfrac{7ab}{a^2b^2+9ab}$
   (e) $\dfrac{xy}{xy+x}$
6. Simplify
   (a) $\dfrac{x-4}{(x-4)(x-2)}$ (b) $\dfrac{2x-4}{x^2+x-6}$
   (c) $\dfrac{3x}{3x^2+6x}$ (d) $\dfrac{x^2+2x+1}{x^2-2x-3}$
   (e) $\dfrac{2(x-3)}{(x-3)^2}$ (f) $\dfrac{x-3}{(x-3)^2}$
   (g) $\dfrac{x-3}{2(x-3)^2}$ (h) $\dfrac{4(x-3)}{2(x-3)^2}$
   (i) $\dfrac{x+4}{2(x+4)^2}$ (j) $\dfrac{x+4}{2(x+4)}$
   (k) $\dfrac{2(x+4)}{(x+4)}$ (l) $\dfrac{(x+4)(x-3)}{x-3}$
   (m) $\dfrac{x+4}{(x-3)(x+4)}$ (n) $\dfrac{x+3}{x^2+7x+12}$
   (o) $\dfrac{x+4}{2x+8}$ (p) $\dfrac{x+4}{2x+9}$

Multiplication and division of algebraic fractions

To multiply two algebraic fractions together we multiply their numerators together and multiply their denominators together:

Key point

$$\frac{a}{b} \times \frac{c}{d} = \frac{a \times c}{b \times d}$$

Any common factors in the result should be cancelled.

WORKED EXAMPLES

9.10 Simplify

$$\frac{4}{5} \times \frac{x}{y}$$

Solution We multiply the numerators together and multiply the denominators together. That is,

$$\frac{4}{5} \times \frac{x}{y} = \frac{4x}{5y}$$

9.11 Simplify

$$\frac{4}{x} \times \frac{3y}{16}$$

Solution The numerators are multiplied together and the denominators are multiplied together. Therefore

$$\frac{4}{x} \times \frac{3y}{16} = \frac{4 \times 3y}{16x}$$

Because $16x = 4 \times 4x$, the common factor 4 can be cancelled. So

$$\frac{4 \times 3y}{16x} = \frac{\not{4} \times 3y}{\not{4} \times 4x} = \frac{3y}{4x}$$

9.12 Simplify

(a) $\frac{1}{2} \times x$ (b) $\frac{1}{2} \times (a + b)$

Solution (a) Writing x as $\frac{x}{1}$ we can state

$$\frac{1}{2} \times x = \frac{1}{2} \times \frac{x}{1} = \frac{1 \times x}{2 \times 1} = \frac{x}{2}$$

(b) Writing $a + b$ as $\frac{a+b}{1}$ we can state

$$\frac{1}{2} \times (a + b) = \frac{1}{2} \times \frac{(a+b)}{1} = \frac{1 \times (a+b)}{2 \times 1} = \frac{a+b}{2}$$

9.13 Simplify

$$\frac{4x^2}{y} \times \frac{3x^3}{yz}$$

Solution We multiply the numerators together and multiply the denominators together:

$$\frac{4x^2}{y} \times \frac{3x^3}{yz} = \frac{4x^2 \times 3x^3}{y \times yz} = \frac{12x^5}{y^2 z}$$

9.14 Simplify

$$5 \times \left(\frac{x-3}{25}\right)$$

Solution This means

$$\frac{5}{1} \times \frac{x-3}{25}$$

which equals

$$\frac{5 \times (x-3)}{1 \times 25}$$

A common factor of 5 can be cancelled from the numerator and denominator to give

$$\frac{(x-3)}{5}$$

9.15 Simplify

$$-\frac{1}{5} \times \frac{3x-4}{8}$$

Solution We can write

$$-\frac{1}{5} \times \frac{3x-4}{8} = -\frac{1 \times (3x-4)}{5 \times 8} = -\frac{3x-4}{40}$$

Note that the answer can also be expressed as $\frac{(4-3x)}{40}$ because

$$-\frac{3x-4}{40} = \frac{-1}{1} \times \frac{3x-4}{40} = \frac{-3x+4}{40} = \frac{4-3x}{40}$$

You should be aware from the last worked example that a solution can often be expressed in a number of equivalent ways.

WORKED EXAMPLES

9.16 Simplify

$$\frac{a}{a+b} \times \frac{b}{5a^2}$$

Solution

$$\frac{a}{a+b} \times \frac{b}{5a^2} = \frac{ab}{5a^2(a+b)}$$

Cancelling the common factor of a in numerator and denominator gives

$$\frac{b}{5a(a+b)}$$

9.17 Simplify

$$\frac{x^2+4x+3}{2x+8} \times \frac{x+4}{x+1}$$

Solution Before multiplying the two fractions together we should try to factorise if possible so that common factors can be identified. By factorising, we can write the given expressions as

$$\frac{(x+1)(x+3)}{2(x+4)} \times \frac{x+4}{x+1} = \frac{(x+1)(x+3)(x+4)}{2(x+4)(x+1)}$$

Cancelling common factors this simplifies to just

$$\frac{x+3}{2}$$

Division is performed by inverting the second fraction and multiplying.

Key point

$$\frac{a}{b} \div \frac{c}{d} = \frac{a}{b} \times \frac{d}{c}$$

WORKED EXAMPLES

9.18 Simplify

$$\frac{10a}{b} \div \frac{a^2}{3b}$$

Solution The second fraction is inverted and then multiplied by the first. That is,

$$\frac{10a}{b} \div \frac{a^2}{3b} = \frac{10a}{b} \times \frac{3b}{a^2} = \frac{30ab}{a^2b} = \frac{30}{a}$$

9.19 Simplify

$$\frac{x^2y^3}{z} \div \frac{y}{x}$$

Solution

$$\frac{x^2y^3}{z} \div \frac{y}{x} = \frac{x^2y^3}{z} \times \frac{x}{y} = \frac{x^3y^3}{zy}$$

Any common factors in the result can be cancelled. So

$$\frac{x^3y^3}{zy} = \frac{x^3y^2}{z}$$

Self-assessment question 9.2

1. The technique of multiplying and dividing algebraic fractions is identical to that used for numbers. True or false?

Exercise 9.2

1. Simplify

(a) $\frac{1}{2} \times \frac{y}{3}$ (b) $\frac{1}{3} \times \frac{z}{2}$ (c) $\frac{2}{5}$ of $\frac{1}{y}$

(d) $\frac{2}{5}$ of $\frac{1}{x}$ (e) $\frac{3}{4}$ of $\frac{x}{y}$ (f) $\frac{3}{5} \times \frac{x^2}{y}$

(g) $\frac{x}{y} \times \frac{3}{5}$ (h) $\frac{7}{8} \times \frac{x}{2y}$ (i) $\frac{1}{2} \times \frac{1}{2x}$

(j) $\frac{1}{2} \times \frac{x}{2}$ (k) $\frac{1}{2} \times \frac{2}{x}$ (l) $\frac{1}{3} \times \frac{x}{3}$

(m) $\frac{1}{3} \times \frac{3}{x}$ (n) $\frac{1}{3} \times \frac{1}{3x}$ (o) $\frac{1}{3} \times \frac{3x}{2}$

2. Simplify

(a) $\frac{1}{2} \div \frac{x}{2}$ (b) $\frac{1}{2} \div \frac{2}{x}$ (c) $\frac{2}{x} \div \frac{2}{x}$

(d) $\frac{x}{2} \div \frac{1}{2}$ (e) $\frac{2}{x} \div 2$ (f) $\frac{2}{x} \div \frac{1}{2}$

(g) $\frac{3}{x} \div \frac{1}{2}$

3. Simplify the following:

(a) $\frac{5}{4} \times \frac{a}{25}$ (b) $\frac{5}{4} \times \frac{a}{b}$ (c) $\frac{8a}{b^2} \times \frac{b}{16a^2}$

(d) $\frac{9x}{3y} \times \frac{2x}{y^2}$ (e) $\frac{3}{5a} \times \frac{b}{a}$ (f) $\frac{1}{4} \times \frac{x}{y}$

(g) $\frac{1}{3} \times \frac{x}{x+y}$ (h) $\frac{x-3}{x+4} \times \frac{1}{3x-9}$

4. Simplify the following:

(a) $\frac{3}{x} \times \frac{xy}{z^3}$ (b) $\frac{(3+x)}{x} \div \frac{y}{x}$ (c) $\frac{4}{3} \div \frac{16}{x}$

(d) $\frac{a}{bc^2} \times \frac{b^2c}{a}$

5. Simplify

(a) $\frac{x+2}{(x+5)(x+4)} \times \frac{x+5}{x+2}$

(b) $\frac{x-2}{4} \div \frac{x}{16}$ (c) $\frac{12ab}{5ef} \div \frac{4ab^2}{f}$

(d) $\frac{x+3y}{2x} \div \frac{y}{4x^2}$ (e) $\frac{3}{x} \times \frac{3}{y} \times \frac{1}{z}$

6. Simplify

$$\frac{1}{x+1} \times \frac{2x+2}{x+3}$$

7. Simplify

$$\frac{x+1}{x+2} \times \frac{x^2+6x+8}{x^2+4x+3}$$

Addition and subtraction of algebraic fractions

For revision of adding and subtracting fractions see pages 66–69.

The method is the same as that for adding or subtracting numerical fractions. Note that it is not correct to simply add or subtract the numerator and denominator. The lowest common denominator must first be found. This is the simplest expression that contains all original denominators as its factors. Each fraction is then written with this common denominator. The fractions can then be added or subtracted by adding or subtracting just the numerators, and dividing the result by the common denominator.

WORKED EXAMPLES

9.20 Add the fractions $\frac{3}{4}$ and $\frac{1}{x}$.

Solution We must find $\frac{3}{4}+\frac{1}{x}$. To do this we must first rewrite the fractions to ensure they have a common denominator. The common denominator is the simplest expression that has the given denominators as its factors. The simplest such expression is $4x$. We write

$$\frac{3}{4} \text{ as } \frac{3x}{4x}, \qquad \text{and} \qquad \frac{1}{x} \text{ as } \frac{4}{4x}$$

Then

$$\frac{3}{4}+\frac{1}{x}=\frac{3x}{4x}+\frac{4}{4x}$$

$$=\frac{3x+4}{4x}$$

No further simplification is possible.

9.21 Simplify

$$\frac{3}{x}+\frac{4}{x^2}$$

Solution The expression $\frac{3}{x}$ is rewritten as $\frac{3x}{x^2}$, which makes the denominators of both terms x^2 but leaves the value of the expression unaltered. Note that both the original denominators, x and x^2, are factors of the new denominator. We call this denominator the **lowest common denominator.** The fractions are then added by adding just the numerators. That is,

$$\frac{3x}{x^2}+\frac{4}{x^2}=\frac{3x+4}{x^2}$$

9.22 Express $\frac{5}{a} - \frac{4}{b}$ as a single fraction.

Solution Both fractions are rewritten to have the same denominator. The simplest expression containing both a and b as its factors is ab. Therefore ab is the lowest common denominator. Then

$$\frac{5}{a} - \frac{4}{b} = \frac{5b}{ab} - \frac{4a}{ab} = \frac{5b - 4a}{ab}$$

9.23 Write $\frac{4}{x+y} - \frac{3}{y}$ as a single fraction.

Solution The simplest expression that contains both denominators as its factors is $(x + y)y$. We must rewrite each term so that it has this denominator.

$$\frac{4}{x+y} = \frac{4}{x+y} \times \frac{y}{y} = \frac{4y}{(x+y)y}$$

Similarly,

$$\frac{3}{y} = \frac{3}{y} \times \frac{x+y}{x+y} = \frac{3(x+y)}{(x+y)y}$$

The fractions are then subtracted by subtracting just the numerators:

$$\frac{4}{x+y} - \frac{3}{y} = \frac{4y}{(x+y)y} - \frac{3(x+y)}{(x+y)y} = \frac{4y - 3(x+y)}{(x+y)y}$$

which simplifies to

$$\frac{y - 3x}{(x+y)y}$$

9.24 Express as a single fraction

$$\frac{2}{x+3} + \frac{5}{x-1}$$

Solution The simplest expression having both $x + 3$ and $x - 1$ as its factors is

$$(x + 3)(x - 1)$$

This is the lowest common denominator. Each term is rewritten so that it has this denominator. Thus

$$\frac{2}{x+3} = \frac{2(x-1)}{(x+3)(x-1)} \quad \text{and} \quad \frac{5}{x-1} = \frac{5(x+3)}{(x+3)(x-1)}$$

Then

$$\frac{2}{x+3}+\frac{5}{x-1}=\frac{2(x-1)}{(x+3)(x-1)}+\frac{5(x+3)}{(x+3)(x-1)}$$

$$=\frac{2(x-1)+5(x+3)}{(x+3)(x-1)}$$

which simplifies to

$$\frac{7x+13}{(x+3)(x-1)}$$

9.25 Express as a single fraction

$$\frac{1}{x-4}+\frac{1}{(x-4)^2}$$

Solution The simplest expression having $x-4$ and $(x-4)^2$ as its factors is $(x-4)^2$. Both fractions are rewritten with this denominator:

$$\frac{1}{x-4}+\frac{1}{(x-4)^2}=\frac{(x-4)}{(x-4)^2}+\frac{1}{(x-4)^2}$$

$$=\frac{x-4+1}{(x-4)^2}$$

$$=\frac{x-3}{(x-4)^2}$$

Self-assessment question 9.3

1. Explain what is meant by the 'lowest common denominator' and how it is found.

Exercise 9.3

1. Express each of the following as a single fraction:

(a) $\frac{z}{2}+\frac{z}{3}$ (b) $\frac{x}{3}+\frac{x}{4}$ (c) $\frac{y}{5}+\frac{y}{25}$

2. Express each of the following as a single fraction:

(a) $\frac{1}{2}+\frac{1}{x}$ (b) $\frac{1}{2}+x$ (c) $\frac{1}{3}+y$

(d) $\frac{1}{3}+\frac{1}{y}$ (e) $8+\frac{1}{y}$

3. Express each of the following as a single fraction:

(a) $\frac{5}{x}-\frac{1}{2}$ (b) $\frac{5}{x}+2$ (c) $\frac{3}{x}-\frac{1}{3}$

(d) $\frac{x}{3}-\frac{1}{2}$ (e) $\frac{3}{x}+\frac{1}{3}$

4. Express each of the following as a single fraction:

(a) $\frac{3}{x}+\frac{4}{y}$ (b) $\frac{3}{x^2}+\frac{4y}{x}$ (c) $\frac{4ab}{x}+\frac{3ab}{2y}$

(d) $\frac{4xy}{a}+\frac{3xy}{2b}$ (e) $\frac{3}{x}-\frac{6}{2x}$ (f) $\frac{3x}{2y}-\frac{7y}{4x}$

(g) $\frac{3}{x+y}-\frac{2}{y}$ (h) $\frac{1}{a+b}-\frac{1}{a-b}$

(i) $2x+\frac{1}{2x}$ (j) $2x-\frac{1}{2x}$

5. Express each of the following as a single fraction:

(a) $\frac{x}{y}+\frac{3x^2}{z}$ (b) $\frac{4}{a}+\frac{5}{b}$

(c) $\frac{6x}{y}-\frac{2y}{x}$ (d) $3x-\frac{3x+1}{4}$

(e) $\frac{5a}{12}+\frac{9a}{18}$ (f) $\frac{x-3}{4}+\frac{3}{5}$

6. Express each of the following as a single fraction:

(a) $\frac{1}{x+1}+\frac{1}{x+2}$ (b) $\frac{1}{x-1}+\frac{2}{x+3}$

(c) $\frac{3}{x+5}+\frac{1}{x+4}$ (d) $\frac{1}{x-2}+\frac{3}{x-4}$

(e) $\frac{3}{2x+1}+\frac{1}{x+1}$ (f) $\frac{3}{1-2x}+\frac{1}{x}$

(g) $\frac{3}{x+1}+\frac{4}{(x+1)^2}$

(h) $\frac{1}{x-1}+\frac{1}{(x-1)^2}$

Test and assignment exercises 9

1. Simplify

(a) $\frac{5a}{4a+3ab}$ (b) $\frac{5a}{30a+15b}$ (c) $\frac{5ab}{30a+15b}$ (d) $\frac{5ab}{ab+7ab}$ (e) $\frac{y}{13y+y^2}$ (f) $\frac{13y+y^2}{y}$

2. Simplify the following:

(a) $\frac{5a}{7}\times\frac{14b}{2}$ (b) $\frac{3}{x}+\frac{7}{3x}$ (c) $t-\frac{4-t}{2}$ (d) $4(x+3)-\frac{(4x-5)}{3}$

(e) $\frac{7}{x}+\frac{3}{2x}+\frac{5}{3x}$ (f) $x+\frac{3x}{y}$ (g) $xy+\frac{1}{xy}$ (h) $\frac{1}{x+y}+\frac{2}{x-y}$

3. Simplify the following:

(a) $\frac{y}{9}+\frac{2y}{7}$ (b) $\frac{3}{x}-\frac{5}{3x}+\frac{4}{5x}$ (c) $\frac{3x}{2y}+\frac{5y}{6x}$ (d) $m+\frac{m+n}{2}$

(e) $m-\frac{m+n}{2}$ (f) $m-\frac{m-n}{2}$ (g) $\frac{3s-5}{10}-\frac{2s-3}{15}$

4. Simplify

$$\frac{x^2-x}{x-1}$$

Chapter 10
Factorisation

Factorisation

Objectives: This chapter:

- explains what is meant by the 'factors' of an algebraic expression
- shows how an algebraic expression can be factorised
- shows how to factorise quadratic expressions

Factors and common factors

A number is **factorised** when it is written as a product. For example, 15 may be factorised into 3×5. We say that 3 and 5 are **factors** of 15. The number 16 can be written as 8×2 or 4×4, or even as 16×1, and so the factorisation may not be unique.

Algebraic expressions can also be factorised. Consider the expression $5x + 20y$. Both $5x$ and $20y$ have the number 5 common to both terms. We say that 5 is a **common factor**. Any common factors can be written outside a bracketed term. Thus $5x + 20y = 5(x + 4y)$. Removal of the brackets will result in the original expression and can always be used to check your answer. We see that factorisation can be thought of as reversing the process of removing brackets. Similarly, if we consider the expression $x^2 + 2x$, we note that both terms contain the factor x, and so $x^2 + 2x$ can be written as $x(x + 2)$. Hence x and $x + 2$ are both factors of $x^2 + 2x$.

WORKED EXAMPLES

10.1 Factorise $3x + 12$.

Solution The number 12 can be factorised as 3×4 so that 3 is a common factor of $3x$ and 12. We can write $3x + 12 = 3x + 3(4)$. Any common factors are written in front of a bracket and the contents of the bracket are adjusted accordingly. So

$$3x + 3(4) = 3(x + 4)$$

Note again that this answer can be checked by removing the brackets.

10.2 **List the ways in which $15x^2$ can be written as a product of its factors.**

Solution $15x^2$ can be written in many different ways. Some of these are $15x^2 \times 1$, $15x \times x$, $15 \times x^2$, $5x \times 3x$, $5 \times 3x^2$, and $3 \times 5x^2$.

10.3 Factorise $8x^2 - 12x$.

Solution We can write $8x^2 - 12x = (4x)(2x) - (4x)3$ so that both terms contain the factor $4x$. This is placed at the front of a bracket to give

$$8x^2 - 12x = 4x(2x - 3)$$

10.4 **What factors are common to the terms $5x^2$ and $15x^3$? Factorise $5x^2 + 15x^3$.**

Solution Both terms contain a factor of 5. Because x^3 can be written $x^2 \times x$, both $5x^2$ and $15x^3$ contain a factor x^2. Therefore

$$5x^2 + 15x^3 = 5x^2 + (5x^2)(3x) = 5x^2(1 + 3x)$$

10.5 Factorise $6x + 3x^2 + 9xy$.

Solution By careful inspection of all of the terms we see that $3x$ is a factor of each term. Hence

$$6x + 3x^2 + 9xy = 3x(2 + x + 3y)$$

Hence the factors of $6x + 3x^2 + 9xy$ are $3x$ and $2 + x + 3y$.

Self-assessment question 10.1

1. Explain what is meant by 'factorising an expression'.

Exercise 10.1

1. Remove the brackets from
   (a) $9(x + 3)$ (b) $-5(x - 2)$
   (c) $\frac{1}{2}(x + 1)$
   (d) $-(a - 3b)$ (e) $\frac{1}{2(x + y)}$
   (f) $\frac{x}{y(x - y)}$

2. List all the factors of each of (a) $4x^2$, (b) $6x^3$.

3. Factorise
   (a) $3x + 18$ (b) $3y - 9$ (c) $-3y - 9$
   (d) $-3 - 9y$ (e) $20 + 5t$ (f) $20 - 5t$
   (g) $-5t - 20$ (h) $3x + 12$ (i) $17t + 34$
   (j) $-36 + 4t$

4. Factorise
   (a) x^4+2x (b) x^4-2x (c) $3x^4-2x$
   (d) $3x^4+2x$ (e) $3x^4+2x^2$
   (f) $3x^4+2x^3$ (g) $17z-z^2$
   (h) $-xy+3x$ (i) $-xy+3y$
   (j) $x+2xy+3xyz$

5. Factorise
   (a) $10x+20y$ (b) $12a+3b$
   (c) $4x-6xy$ (d) $7a+14$
   (e) $10m-15$ (f) $\dfrac{1}{5a+35b}$
   (g) $\dfrac{1}{5a^2+35ab}$

6. Factorise
   (a) $15x^2+3x$ (b) $4x^2-3x$
   (c) $4x^2-8x$ (d) $15-3x^2$
   (e) $10x^3+5x^2+15x^2y$
   (f) $6a^2b-12ab^2$
   (g) $16abc-8ab^2+24bc$

Factorising quadratic expressions

Expressions of the form ax^2+bx+c, where a, b, and c are numbers, are called **quadratic expressions**. The numbers b or c may equal zero but a must not be zero. The number a is called the **coefficient** of x^2, b is the coefficient of x, and c is called the **constant term**.

We see that

$$2x^2+3x-1, \quad x^2+3x+2, \quad x^2+7 \quad \text{and} \quad 2x^2-x$$

are all quadratic expressions.

Key point

An expression of the form ax^2+bx+c, where a, b and c are numbers, is called a quadratic expression. The coefficient of x^2 is a, the coefficient of x is b, and the constant term is c.

To factorise such an expression means to express it as a product of two terms. For example, removing the brackets from $(x+6)(x-3)$ gives $x^2+3x-18$ (see Worked Example 8.22). Reversing the process, $x^2+3x-18$ can be factorised to $(x+6)(x-3)$. Not all quadratic expressions can be factorised in this way. We shall now explore how such factorisation is attempted.

Quadratic expressions where the coefficient of x^2 is 1

Consider the expression $(x+m)(x+n)$. Removing the brackets we find

$$\begin{aligned}(x+m)(x+n) &= (x+m)x+(x+m)n\\ &= x^2+mx+nx+mn\\ &= x^2+(m+n)x+mn\end{aligned}$$

Note that the coefficient of the x term is the sum $m+n$ and the constant term is the product mn. Using this information several quadratic expressions can be factorised by careful inspection. For example, suppose we wish to factorise x^2+5x+6. We know that $x^2+(m+n)x+mn$ can be factorised to $(x+m)(x+n)$. We seek values of m and n so that

$$x^2+5x+6=x^2+(m+n)x+mn$$

Comparing the coefficients of x on both sides we require

$$5=m+n$$

Comparing the constant terms on both sides we require

$$6=mn$$

By inspection we see that $m=3$ and $n=2$ have this property and so

$$x^2+5x+6=(x+3)(x+2)$$

Note that the answer can be easily checked by removing the brackets again.

WORKED EXAMPLES

10.6 Factorise the quadratic expression $x^2+8x+12$.

Solution The factorisation of $x^2+8x+12$ will be of the form $(x+m)(x+n)$. This means that mn must equal 12 and $m+n$ must equal 8. The two numbers must therefore be 2 and 6. So

$$x^2+8x+12=(x+2)(x+6)$$

Note again that the answer can be checked by removing the brackets.

10.7 Factorise $x^2+10x+25$.

Solution We try to factorise in the form $(x+m)(x+n)$. We require $m+n$ to equal 10 and mn to equal 25. If $m=5$ and $n=5$ this requirement is met. Therefore $x^2+10x+25=(x+5)(x+5)$. It is usual practice to write this as $(x+5)^2$.

10.8 Factorise x^2-121.

Solution In this example the x term is missing. We still attempt to factorise as $(x+m)(x+n)$. We require $m+n$ to equal 0 and mn to equal -121. Some thought shows that if $m=11$ and $n=-11$ this requirement is met. Therefore $x^2-121=(x+11)(x-11)$.

10.9 Factorise x^2-5x+6.

Solution We try to factorise in the form $(x+m)(x+n)$. We require $m+n$ to equal -5 and mn to equal 6. By inspection we see that if $m=-3$ and $n=-2$ this requirement is met. Therefore $x^2-5x+6=(x-3)(x-2)$.

Quadratic expressions where the coefficient of x^2 is not 1

These expressions are a little harder to factorise. All possible factors of the first and last terms must be found, and various combinations of these should be attempted until the required answer is found. This involves trial and error along with educated guesswork and practice.

WORKED EXAMPLES

10.10 Factorise, if possible, the expression $2x^2 + 11x + 12$.

Solution The factors of the first term, $2x^2$, are $2x$ and x. The factors of the last term, 12, are

$$12, 1 \quad -12, -1 \quad 6, 2 \quad -6, -2 \quad \text{and} \quad 4, 3 \quad -4, -3$$

We can try each of these combinations in turn to find which gives us a coefficient of x of 11. For example, removing the brackets from

$$(2x + 12)(x + 1)$$

gives

$$\begin{aligned}(2x + 12)(x + 1) &= (2x + 12)x + (2x + 12)(1) \\ &= 2x^2 + 12x + 2x + 12 \\ &= 2x^2 + 14x + 12\end{aligned}$$

which has an incorrect middle term. By trying further combinations it turns out that the only one producing a middle term of $11x$ is $(2x + 3)(x + 4)$ because

$$\begin{aligned}(2x + 3)(x + 4) &= (2x + 3)(x) + (2x + 3)(4) \\ &= 2x^2 + 3x + 8x + 12 \\ &= 2x^2 + 11x + 12\end{aligned}$$

so that $(2x + 3)(x + 4)$ is the correct factorisation.

10.11 Factorise, if possible, $4x^2 + 6x + 2$.

Solution Before we try to factorise this quadratic expression notice that there is a factor of 2 in each term so that we can write it as $2(2x^2 + 3x + 1)$. Now consider the quadratic expression $2x^2 + 3x + 1$. The factors of the first term, $2x^2$, are $2x$ and x. The factors of the last term, 1, are simply 1 and 1, or -1 and -1. We can try these combinations in turn to find which gives us a middle term of $3x$. Removing the brackets from $(2x + 1)(x + 1)$ gives $2x^2 + 3x + 1$, which has the correct middle term. Finally, we can write

$$4x^2 + 6x + 2 = 2(2x^2 + 3x + 1) = 2(2x + 1)(x + 1)$$

10.12 Factorise $6x^2 + 7x - 3$.

Solution The first term may be factorised as $6x \times x$ and also as $3x \times 2x$. The factors of the last term are

$$3, -1 \quad \text{and} \quad -3, 1$$

We need to try each combination in turn to find which gives us a coefficient of x of 7. For example, removing the brackets from

$$(6x + 3)(x - 1)$$

gives

$$\begin{aligned}(6x + 3)(x - 1) &= (6x + 3)x + (6x + 3)(-1)\\ &= 6x^2 + 3x - 6x - 3\\ &= 6x^2 - 3x - 3\end{aligned}$$

which has an incorrect middle term. By trying further combinations it turns out that the only one producing a middle term of $7x$ is $(3x - 1)(2x + 3)$ because

$$\begin{aligned}(3x - 1)(2x + 3) &= (3x - 1)2x + (3x - 1)(3)\\ &= 6x^2 - 2x + 9x - 3\\ &= 6x^2 + 7x - 3\end{aligned}$$

The correct factorisation is therefore $(3x - 1)(2x + 3)$.

Until you have sufficient experience at factorising quadratic expressions you must be prepared to go through the process of trying all possible combinations until the correct answer is found.

Self-assessment question 10.2

1. Not all quadratic expressions can be factorised. Try to find an example of one such expression.

Exercise 10.2

1. Factorise the following quadratic expressions:
   (a) $x^2 + 3x + 2$ (b) $x^2 + 13x + 42$
   (c) $x^2 + 2x - 15$ (d) $x^2 + 9x - 10$
   (e) $x^2 - 11x + 24$ (f) $x^2 - 100$
   (g) $x^2 + 4x + 4$ (h) $x^2 - 36$
   (i) $x^2 - 25$ (j) $x^2 + 10x + 9$
   (k) $x^2 + 8x - 9$ (l) $x^2 - 8x - 9$
   (m) $x^2 - 10x + 9$ (n) $x^2 - 5x$

2. Factorise the following quadratic expressions:
   (a) $2x^2 - 5x - 3$ (b) $3x^2 - 5x - 2$
   (c) $10x^2 + 11x + 3$ (d) $2x^2 + 12x + 16$
   (e) $2x^2 + 5x + 3$ (f) $3s^2 + 5s + 2$
   (g) $3z^2 + 17z + 10$ (h) $9x^2 - 36$
   (i) $4x^2 - 25$

3. (a) By removing the brackets show that

$$(x+y)(x-y)=x^2-y^2$$

This result is known as the **difference of two squares**.

(b) Using the result in part (a) write down the factorisation of
(i) $16x^2-1$ (ii) $16x^2-9$
(iii) $25t^2-16r^2$

4. Factorise the following quadratic expressions:
(a) $x^2+3x-10$ (b) $2x^2-3x-20$
(c) $9x^2-1$ (d) $10x^2+14x-12$
(e) $x^2+15x+26$ (f) $-x^2-2x+3$

5. Factorise
(a) $100-49x^2$ (b) $36x^2-25y^2$
(c) $\frac{1}{4}-9v^2$ (d) $\frac{x^2}{y^2}-4$

Test and assignment exercises 10

1. Factorise the following expressions:
(a) $7x+49$ (b) $121x+22y$ (c) a^2+ab (d) $ab+b^2$ (e) ab^2+ba^2

2. Factorise the following quadratic expressions:
(a) $3x^2+x-2$ (b) x^2-144 (c) s^2-5s+6 (d) $2y^2-y-15$

3. Factorise the following:
(a) $1-x^2$ (b) x^2-1 (c) $9-x^2$ (d) x^2-81 (e) $25-y^2$

4. Factorise the denominators of the following expressions:
(a) $\frac{1}{x^2+6x}$ (b) $\frac{3}{s^2+3s+2}$ (c) $\frac{3}{s^2+s-2}$ (d) $\frac{5}{x^2+11x+28}$ (e) $\frac{x}{2x^2-17x-9}$

Chapter 11
Transposing Formulae

03

Transposing formulae

Objectives: This chapter:

- explains how formulae can be rearranged or transposed

Rearranging a formula

In the formula for the area of a circle, $A = \pi r^2$, we say that A is the **subject** of the formula. The subject appears by itself on one side, usually the left, of the formula, and nowhere else. If we are asked to **transpose** the formula for r, then we must rearrange the formula so that r becomes the subject. The rules for transposing formulae are quite simple. Essentially whatever you do to one side you must also do to the whole of the other side.

Key point

To transpose a formula you may:

- add the same quantity to both sides
- subtract the same quantity from both sides
- multiply or divide both sides by the same quantity
- perform operations on both sides, such as 'square both sides', 'square root both sides' etc.

WORKED EXAMPLES

11.1 The circumference C of a circle is given by the formula $C = 2\pi r$. Transpose this to make r the subject.

Solution The intention is to obtain r by itself on the left-hand side. Starting with $C = 2\pi r$ we must try to isolate r. Dividing both sides by 2π we find

$$C = 2\pi r$$

$$\frac{C}{2\pi} = \frac{2\pi r}{2\pi}$$

$$\frac{C}{2\pi} = r \qquad \text{by cancelling the common factor } 2\pi \text{ on the right}$$

Finally we can write $r = \frac{C}{2\pi}$ and the formula has been transposed for r.

11.2 **Transpose the formula $y = 3(x + 7)$ for x.**

Solution We must try to obtain x on its own on the left-hand side. This can be done in stages. Dividing both sides by 3 we find

$$\frac{y}{3} = \frac{3(x+7)}{3} = x + 7 \qquad \text{by cancelling the common factor 3}$$

Subtracting 7 from both sides then gives

$$\frac{y}{3} - 7 = x + 7 - 7$$

so that

$$\frac{y}{3} - 7 = x$$

Equivalently we can write $x = \dfrac{y}{3} - 7$, and the formula has been transposed for x.

Alternatively, again starting from $y = 3(x + 7)$ we could proceed as follows. Removing the brackets we obtain $y = 3x + 21$. Then, subtracting 21 from both sides,

$$y - 21 = 3x + 21 - 21 = 3x$$

Dividing both sides by 3 will give x on its own:

$$\frac{y-21}{3} = x \qquad \text{so that} \qquad x = \frac{y-21}{3}$$

Noting that

$$\frac{y-21}{3} = \frac{y}{3} - \frac{21}{3} = \frac{y}{3} - 7$$

we see that this answer is equivalent to the expression obtained previously.

11.3 Transpose the formula $y - z = 3(x + 2)$ for x.

Solution Dividing both sides by 3 we find

$$\frac{y - z}{3} = x + 2$$

Subtracting 2 from both sides gives

$$\frac{y - z}{3} - 2 = x + 2 - 2$$
$$= x$$

so that

$$x = \frac{y - z}{3} - 2$$

11.4 Transpose $x + xy = 7$, (a) for x, (b) for y.

Solution (a) In the expression $x + xy$, note that x is a common factor. We can then write the given formula as $x(1 + y) = 7$. Dividing both sides by $1 + y$ will isolate x. That is,

$$\frac{x(1 + y)}{1 + y} = \frac{7}{1 + y}$$

Cancelling the common factor $1 + y$ on the left-hand side gives

$$x = \frac{7}{1 + y}$$

and the formula has been transposed for x.

(b) To transpose $x + xy = 7$ for y we first subtract x from both sides to give $xy = 7 - x$. Finally dividing both sides by x gives

$$\frac{xy}{x} = \frac{7 - x}{x}$$

Cancelling the common factor of x on the left-hand side gives $y = \dfrac{7 - x}{x}$, which is the required transposition.

11.5 Transpose the formula $I = \frac{V}{R}$ for R.

Solution Starting with $I = \frac{V}{R}$ we first multiply both sides by R. This gives

$$IR = \frac{V}{R} \times R = V$$

Then, dividing both sides by I gives

$$\frac{IR}{I} = \frac{V}{I}$$

so that $R = \frac{V}{I}$.

11.6 **Make x the subject of the formula $y = \dfrac{4}{x-7}$**

Solution We must try to obtain x on its own on the left-hand side. It is often useful to multiply both sides of the formula by the same quantity in order to remove fractions. Multiplying both sides by $(x - 7)$ we find

$$(x-7)y = (x-7) \times \frac{4}{x-7} = 4$$

Removing the brackets on the left-hand side we find

$$xy - 7y = 4 \qquad \text{so that} \qquad xy = 7y + 4$$

Finally, dividing both sides by y we obtain

$$\frac{xy}{y} = \frac{7y+4}{y}$$

That is,

$$x = \frac{7y+4}{y}$$

11.7 **Transpose the formula**

$$T = 2\pi\sqrt{\frac{l}{g}}$$

to make l the subject.

Solution We must attempt to isolate l. We can do this in stages as follows. Dividing both sides by 2π we find

$$\frac{T}{2\pi} = \frac{2\pi\sqrt{\frac{l}{g}}}{2\pi} = \sqrt{\frac{l}{g}} \qquad \text{by cancelling the common factor } 2\pi$$

Recall from §7.3 that $(\sqrt{a})^2 = a$.

The square root sign over the l/g can be removed by squaring both sides. Recall that if a term containing a square root is squared then the square root will disappear.

$$\left(\frac{T}{2\pi}\right)^2 = \frac{l}{g}$$

Then multiplying both sides by g we find

$$g\left(\frac{T}{2\pi}\right)^2 = l$$

Equivalently we have

$$l = g\left(\frac{T}{2\pi}\right)^2$$

and the formula has been transposed for l.

Self-assessment questions 11.1

1. Explain what is meant by the subject of a formula.
2. In what circumstances might you want to transpose a formula?
3. Explain what processes are allowed when transposing a formula.

Exercise 11.1

1. Transpose each of the following formulae to make x the subject:

   (a) $y = 3x$ (b) $y = \frac{1}{x}$ (c) $y = 7x - 5$

   (d) $y = \frac{1}{2}x - 7$ (e) $y = \frac{1}{2x}$

   (f) $y = \frac{1}{2x+1}$ (g) $y = \frac{1}{2x} + 1$

   (h) $y = 18x - 21$ (i) $y = 19 - 8x$

2. Transpose $y = mx + c$, (a) for m, (b) for x, (c) for c.

3. Transpose the following formulae for x:

   (a) $y = 13(x - 2)$ (b) $y = x\left(1 + \frac{1}{x}\right)$

   (c) $y = a + t(x - 3)$

4. Transpose each of the following formulae to make the given variable the subject:

   (a) $y = 7x + 11$, for x (b) $V = IR$, for I

   (c) $V = \frac{4}{3}\pi r^3$, for r (d) $F = ma$, for m

5. Make n the subject of the formula $l = a + (n - 1)d$.

6. If $m = n + t\sqrt{x}$, find an expression for x.

7. Make x the subject of the following formulae:
(a) $y = 1 - x^2$ (b) $y = \dfrac{1}{1 - x^2}$
(c) $y = \dfrac{1 - x^2}{1 + x^2}$

Test and assignment exercises 11

1. Make x the subject of the following formulae:
(a) $y = 13x - 18$ (b) $y = \dfrac{13}{x + 18}$ (c) $y = \dfrac{13}{x} + 18$

2. Transpose the following formulae:
(a) $y = \dfrac{7 + x}{14}$ for x (b) $E = \dfrac{1}{2}mv^2$ for v (c) $8x - 13y = 12$ for y (d) $y = \dfrac{x^2}{2g}$ for x

3. Make x the subject of the formula $v = k/\sqrt{x}$.

4. Transpose the formula $V = \pi r^2 h$ for h.

5. Make r the subject of the formula $y = H + Cr$.

6. Transpose the formula $s = ut + \dfrac{1}{2}at^2$ for a.

7. Transpose the formula $v^2 = u^2 + 2as$ for u.

8. Transpose the formula $k = pv^3$ for v.

Transposing biological formulae

There are many areas of biology which involve transposing, or rearranging, formulae. You will have already come across this in previous chapters, for example those on moles and molarity, and measurements.

Sometimes rearranging formulae involves substituting one formula for another. You have seen an example of this in the moles and molarity chapter and you will also come across it in the chapter on logarithms. Using the example from the moles and molarity chapter, the following equation:

$$number\ of\ moles(mol) = \frac{mass(g)}{molar\ mass(g\ mol^{-1})}$$

is used to substitute the *number of moles* part of this equation:

$$molarity\,(mol^{-1}) = \frac{number\ of\ moles\,(mol)}{volume\,(l)}$$

So instead of *number of moles*, *mass* goes on the top of the equation and *volume* goes on the bottom:

$$molarity\,(mol\ l^{-1}) = \frac{mass(g)}{molar\ mass\,(g\ mol^{-1}) \times volume\,(l)}$$

By substituting one equation for another you can simplify a series of calculations by reducing the number of steps involved.

If you have a problem to solve where you are asked the volume of a sphere, and given its circumference, you could, as demonstrated in the measurements chapter, tackle it as a two stage problem; first determine the radius and then from that, determine the volume. Alternatively you could combine the two equations:

$C = 2\pi r$ or $r = \dfrac{C}{2\pi}$ can be substituted into $V = \dfrac{4}{3}\pi r^3$

to give $V = \dfrac{4}{3}\pi\left(\dfrac{C}{2\pi}\right)^3$ expanding the brackets: $V = \dfrac{4}{3}\pi\dfrac{C^3}{8\pi^3}$ and simplifying: $V = \dfrac{C^3}{6\pi^2}$

This means that you do not have to calculate the radius at all, but can go straight from the circumference to the volume.

One commonly used formula is the Michaelis-Menten equation (see chapter on Graphs and their Equations).

$$v = \frac{V_{max}[S]}{K_m + [S]}$$

This equation is transposed so that $^1/_v$ becomes the subject, and in this form it is referred to as the Lineweaver-Burk plot. The equation is transposed in the following way.

First the equation is turned upside down.

$$\frac{1}{v} = \frac{K_m + [S]}{V_{max}[S]}$$

The right hand side of the equation can be seen as two fractions over a common denominator, so can be split into two fractions.

$$\frac{1}{v} = \frac{K_m}{V_{max}[S]} + \frac{[S]}{V_{max}[S]}$$

In the second fraction the [S] term cancels out.

$$\frac{1}{v} = \frac{K_m}{V_{max}[S]} + \frac{1}{V_{max}}$$

The first fraction is then split into two separate fractions.

$$\frac{1}{v} = \frac{K_m}{V_{max}} \times \frac{1}{[S]} + \frac{1}{V_{max}}$$

This equation is now in the form recognised as the Lineweaver-Burk plot.

There are two other transpositions of the Michaelis-Menten equation.

The Eadie-Hofstee plot $$v = -K_m \frac{v}{[S]} + V_{max}$$

and the Hanes-Woolf plot $$\frac{[S]}{v} = \frac{[S]}{V_{max}} + \frac{K_m}{V_{max}}$$

Test your understanding

1. Rearrange to make mass the subject of the following:

a)
$$\text{moles (mol)} = \frac{\text{mass (g)}}{\text{molar mass}\left(\text{g mol}^{-1}\right)}$$

b)
$$\text{molarity (mol l}^{-1}\text{)} = \frac{\text{mass (g)}}{\text{molar mass (g mol}^{-1}\text{)} \times \text{volume (l)}}$$

2. Rearrange $A = \alpha lc$ to make c the subject:

3. Rearrange to make r the subject of the equation:

a) $C = 2\pi r$

b) $A = \pi r^2$

c) $V = \frac{4}{3}\pi r^3$

4. Take the following formula $v = \frac{V_{\max}[S]}{K_m + [S]}$

and make the following the subject:

a) $V_{\max}$

b) K_m

5. Combine the formula for the circumference of a circle with the formula for the area of a circle, so that you can calculate the area directly from the circumference.

$C = 2\pi r$ $A = \pi r^2$

Chapter 12
Graphs and their Equations

03

Solving equations

Objectives: This chapter:

- explains what is meant by an equation and its solution
- shows how to solve linear, simultaneous and quadratic equations

An **equation** states that two quantities are equal, and will always contain an **unknown quantity** that we wish to find. For example, in the equation $5x + 10 = 20$ the unknown quantity is x. To **solve** an equation means to find all values of the unknown quantity that can be substituted into the equation so that the left side equals the right side. Each such value is called a **solution** or alternatively a **root** of the equation. In the example above the solution is $x = 2$ because when $x = 2$ is substituted both the left side and the right side equal 20. The value $x = 2$ is said to **satisfy** the equation.

Solving linear equations

A **linear** equation is one of the form $ax + b = 0$ where a and b are numbers and the unknown quantity is x. The number a is called the **coefficient** of x. The number b is called the **constant term**. For example, $3x + 7 = 0$ is a linear equation. The coefficient of x is 3 and the constant term is 7. Similarly, $-2x + 17.5 = 0$ is a linear equation. The coefficient of x is -2 and the constant term is 17.5. Note that the unknown quantity occurs only to the first power, that is, as x, and not as x^2, x^3, $x^{1/2}$ etc. Linear equations may appear in other forms that may seem to be different but are nevertheless equivalent. Thus

$$4x + 13 = -7, \qquad 3 - 14x = 0 \qquad \text{and} \qquad 3x + 7 = 2x - 4$$

are all linear equations that could be written in the form $ax + b = 0$ if necessary.

An equation such as $3x^2 + 2 = 0$ is not linear because the unknown quantity occurs to the power 2. Linear equations are solved by trying to obtain the unknown quantity on its own on the left-hand side, that is, by making the unknown quantity the subject of the equation. This is done using the rules given for transposing formulae in Chapter 11. Consider the following examples.

WORKED EXAMPLE

12.1 Solve the equation $x + 10 = 0$.

Solution We make x the subject by subtracting 10 from both sides to give $x = -10$. Therefore $x = -10$ is the solution. It can be easily checked by substituting into the original equation:

$$(-10) + 10 = 0 \qquad \text{as required}$$

Note from the last worked example that the solution should be checked by substitution to ensure it satisfies the given equation. If it does not then a mistake has been made.

WORKED EXAMPLES

12.2 Solve the equation $4x + 8 = 0$.

Solution In order to find the unknown quantity x we attempt to make it the subject of the equation. Subtracting 8 from both sides we find

$$4x + 8 - 8 = 0 - 8 = -8$$

That is,

$$4x = -8$$

Then dividing both sides by 4 gives

$$\frac{4x}{4} = \frac{-8}{4}$$

so that $x = -2$. The solution of the equation $4x + 8 = 0$ is $x = -2$.

12.3 Solve the equation $5x + 17 = 4x - 3$.

Solution First we collect all terms involving x together. This is done by subtracting $4x$ from both sides to remove this term from the right. This gives

$$5x - 4x + 17 = -3$$

That is, $x + 17 = -3$. To make x the subject of this equation we now subtract 17 from both sides to give $x = -3 - 17 = -20$. The solution of the equation $5x + 17 = 4x - 3$ is $x = -20$. Note that the answer can be easily checked by substituting $x = -20$ into the original equation and verifying that the left side equals the right side.

12.4 Solve the equation $\dfrac{x-3}{4} = 1$.

Solution We attempt to obtain x on its own. First note that if we multiply both sides of the equation by 4 this will remove the 4 in the denominator. That is,

$$4 \times \left(\frac{x-3}{4}\right) = 4 \times 1$$

so that

$$x - 3 = 4$$

Finally adding 3 to both sides gives $x = 7$.

Self-assessment questions 12.1

1. Explain what is meant by a root of an equation.
2. Explain what is meant by a linear equation.
3. State the rules that can be used to solve a linear equation.
4. You may think that a formula and an equation look very similar. Try to explain the distinction between a formula and an equation.

Exercise 12.1

1. Verify that the given values of x satisfy the given equations:
   (a) $x = 7$ satisfies $3x + 4 = 25$
   (b) $x = -5$ satisfies $2x - 11 = -21$
   (c) $x = -4$ satisfies $-x - 8 = -4$
   (d) $x = \frac{1}{2}$ satisfies $8x + 4 = 8$
   (e) $x = -\frac{1}{3}$ satisfies $27x + 8 = -1$
   (f) $x = 4$ satisfies $3x + 2 = 7x - 14$
2. Solve the following linear equations:
   (a) $3x = 9$ (b) $\dfrac{x}{3} = 9$ (c) $3t + 6 = 0$
   (d) $3x - 13 = 2x + 9$ (e) $3x + 17 = 21$
   (f) $4x - 20 = 3x + 16$
   (g) $5 - 2x = 2 + 3x$
   (h) $\dfrac{x+3}{2} = 3$ (i) $\dfrac{3x+2}{2} + 3x = 1$
3. Solve the following equations:
   (a) $5(x + 2) = 13$
   (b) $3(x - 7) = 2(x + 1)$
   (c) $5(1 - 2x) = 2(4 - 2x)$
4. Solve the following equations:
   (a) $3t + 7 = 4t - 2$
   (b) $3v = 17 - 4v$
   (c) $3s + 2 = 14(s - 1)$

5. Solve the following linear equations:
   (a) $5t+7=22$ (b) $7-4t=-13$
   (c) $7-4t=27$ (d) $5=14-3t$
   (e) $4x+13=-x+25$
   (f) $\frac{x+3}{2}=\frac{x-3}{4}$ (g) $\frac{1}{3}x+6=\frac{1}{2}x+2$
   (h) $\frac{1}{5}x+7=\frac{1}{3}x+5$
   (i) $\frac{2x+4}{5}=\frac{x-3}{2}$ (j) $\frac{x-7}{8}=\frac{3x+1}{5}$
6. The following equations may not appear to be linear at first sight but they can all be rewritten in the standard form of a linear equation. Find the solution of each equation.
   (a) $\frac{1}{x}=5$ (b) $\frac{1}{x}=\frac{5}{2}$ (c) $\frac{1}{x+1}=\frac{5}{2}$
   (d) $\frac{1}{x-1}=\frac{5}{2}$ (e) $\frac{1}{x}=\frac{1}{2x+1}$
   (f) $\frac{3}{x}=\frac{1}{2x+1}$ (g) $\frac{1}{x+1}=\frac{1}{3x+2}$
   (h) $\frac{3}{x+1}=\frac{2}{4x+1}$

Solving simultaneous equations

Sometimes equations contain more than one unknown quantity. When this happens there are usually two or more equations. For example in the two equations

$$x+2y=14 \qquad 3x+y=17$$

the unknowns are x and y. Such equations are called **simultaneous equations** and to solve them we must find values of x and y that satisfy both equations at the same time. If we substitute $x=4$ and $y=5$ into either of the two equations above we see that the equation is satisfied. We shall demonstrate how simultaneous equations can be solved by removing, or **eliminating**, one of the unknowns.

WORKED EXAMPLES

12.5 Solve the simultaneous equations

$$x+3y=14 \qquad (12.1)$$

$$2x-3y=-8 \qquad (12.2)$$

Solution Note that if these two equations are added the unknown y is removed or eliminated:

$$\begin{array}{rl} x+3y=14 & \\ 2x-3y=-8 & + \\ \hline 3x \quad =6 & \\ \hline \end{array}$$

so that $x = 2$. To find y we substitute $x = 2$ into either equation. Substituting into Equation 12.1 gives

$$2 + 3y = 14$$

Solving this linear equation will give y. We have

$$2 + 3y = 14$$

$$3y = 14 - 2 = 12$$

$$y = \frac{12}{3} = 4$$

Therefore the solution of the simultaneous equations is $x = 2$ and $y = 4$. Note that these solutions should be checked by substituting back into both given equations to check that the left-hand side equals the right-hand side.

12.6 Solve the simultaneous equations

$$5x + 4y = 7 \quad (12.3)$$

$$3x - y = 11 \quad (12.4)$$

Solution Note that, in this example, if we multiply the first equation by 3 and the second by 5 we shall have the same coefficient of x in both equations. This gives

$$15x + 12y = 21 \quad (12.5)$$

$$15x - 5y = 55 \quad (12.6)$$

We can now eliminate x by subtracting Equation 12.6 from Equation 12.5, giving

$$\begin{aligned} 15x + 12y &= 21 \\ 15x - 5y &= 55 \\ \hline 17y &= -34 \end{aligned}$$

from which $y = -2$. In order to find x we substitute our solution for y into either of the given equations. Substituting into Equation 12.3 gives

$$5x + 4(-2) = 7 \quad \text{so that} \quad 5x = 15 \quad \text{or} \quad x = 3$$

The solution of the simultaneous equations is therefore $x = 3$, $y = -2$.

Exercise 12.2

1. Verify that the given values of x and y satisfy the given simultaneous equations.
   (a) $x = 7$, $y = 1$ satisfy $2x - 3y = 11$, $3x + y = 22$
   (b) $x = -7$, $y = 2$ satisfy $2x + y = -12$, $x - 5y = -17$
   (c) $x = -1$, $y = -1$ satisfy $7x - y = -6$, $x - y = 0$

2. Solve the following pairs of simultaneous equations:
   (a) $3x + y = 1, \quad 2x - y = 2$
   (b) $4x + 5y = 21, \quad 3x + 5y = 17$
   (c) $2x - y = 17, \quad x + 3y = 12$
   (d) $-2x + y = -21, \quad x + 3y = -14$
   (e) $-x + y = -10, \quad 3x + 7y = 20$
   (f) $4x - 2y = 2, \quad 3x - y = 4$
3. Solve the following simultaneous equations:
   (a) $5x + y = 36$, $3x - y = 20$
   (b) $x - 3y = -13$, $4x + 2y = -24$
   (c) $3x + y = 30$, $-5x + 3y = -50$
   (d) $3x - y = -5$, $-7x + 3y = 15$
   (e) $11x + 13y = -24$, $x + y = -2$

Solving quadratic equations

A **quadratic equation** is an equation of the form $ax^2 + bx + c = 0$ where a, b and c are numbers and x is the unknown quantity we wish to find. The number a is the **coefficient** of x^2, b is the coefficient of x, and c is the **constant term**. Sometimes b or c may be zero, although a can never be zero. For example

$$x^2 + 7x + 2 = 0 \qquad 3x^2 - 2 = 0 \qquad -2x^2 + 3x = 0 \qquad 8x^2 = 0$$

are all quadratic equations.

Key point

A quadratic equation has the form $ax^2 + bx + c = 0$ where a, b and c are numbers, and x represents the unknown we wish to find.

WORKED EXAMPLES

12.7 State the coefficient of x^2 and the coefficient of x in the following quadratic equations:

(a) $4x^2 + 3x - 2 = 0$ (b) $x^2 - 23x + 17 = 0$ (c) $-x^2 + 19 = 0$

Solution

(a) In the equation $4x^2 + 3x - 2 = 0$ the coefficient of x^2 is 4 and the coefficient of x is 3.

(b) In the equation $x^2 - 23x + 17 = 0$ the coefficient of x^2 is 1 and the coefficient of x is -23.

(c) In the equation $-x^2 + 19 = 0$ the coefficient of x^2 is -1 and the coefficient of x is zero, since there is no term involving just x.

12.8 Verify that both $x = -7$ and $x = 5$ satisfy the quadratic equation $x^2 + 2x - 35 = 0$.

Solution We substitute $x = -7$ into the left-hand side of the equation. This yields

$$(-7)^2 + 2(-7) - 35$$

That is,

$$49 - 14 - 35$$

This simplifies to zero, and so the left-hand side equals the right-hand side of the given equation. Therefore $x = -7$ is a solution.

Similarly, if $x = 5$ we find

$$(5^2) + 2(5) - 35 = 25 + 10 - 35$$

which also simplifies to zero. We conclude that $x = 5$ is also a solution.

Solution by factorisation

If the quadratic expression on the left-hand side of the equation can be factorised solutions can be found using the method in the following examples.

WORKED EXAMPLES

12.9 **Solve the quadratic equation $x^2 + 3x - 10 = 0$.**

Solution The left-hand side of the equation can be factorised to give

$$x^2 + 3x - 10 = (x + 5)(x - 2) = 0$$

Whenever the product of two quantities equals zero, then one or both of these quantities must be zero. It follows that either $x + 5 = 0$ or $x - 2 = 0$ from which $x = -5$ and $x = 2$ are the required solutions.

12.10 **Solve the quadratic equation $6x^2 + 5x - 4 = 0$.**

Solution The left-hand side of the equation can be factorised to give

$$6x^2 + 5x - 4 = (2x - 1)(3x + 4) = 0$$

It follows that either $2x - 1 = 0$ or $3x + 4 = 0$ from which $x = \frac{1}{2}$ and $x = -\frac{4}{3}$ are the required solutions.

12.11 **Solve the quadratic equation $x^2 - 8x = 0$.**

Solution Factorising the left-hand side gives

$$x^2 - 8x = x(x - 8) = 0$$

from which either $x = 0$ or $x - 8 = 0$. The solutions are therefore $x = 0$ and $x = 8$.

12.12 **Solve the quadratic equation $x^2 - 36 = 0$.**

Solution The left-hand side factorises to give

$$x^2 - 36 = (x+6)(x-6) = 0$$

so that $x+6=0$ or $x-6=0$. The solutions are therefore $x=-6$ and $x=6$.

12.13 Solve the equation $x^2 = 81$.

Solution Writing this in the standard form of a quadratic equation we obtain $x^2 - 81 = 0$. The left-hand side can be factorised to give

$$x^2 - 81 = (x-9)(x+9) = 0$$

from which $x-9=0$ or $x+9=0$. The solutions are then $x=9$ and $x=-9$.

Solution of quadratic equations using the formula

When it is difficult or impossible to factorise the quadratic expression ax^2+bx+c, solutions of a quadratic equation can be sought using the following formula:

Key point

If $ax^2+bx+c=0$, then $x = \dfrac{-b \pm \sqrt{b^2-4ac}}{2a}$

This formula gives possibly two solutions: one solution is obtained by taking the positive square root and the second solution by taking the negative square root.

WORKED EXAMPLES

12.14 Solve the equation $x^2+9x+20=0$ using the formula.

Solution Comparing the given equation with the standard form $ax^2+bx+c=0$ we see that $a=1$, $b=9$ and $c=20$. These values are substituted into the formula:

$$x = \frac{-b \pm \sqrt{b^2-4ac}}{2a}$$

$$= \frac{-9 \pm \sqrt{81-4(1)(20)}}{(2)(1)}$$

$$= \frac{-9 \pm \sqrt{81 - 80}}{2}$$

$$= \frac{-9 \pm \sqrt{1}}{2}$$

$$= \frac{-9 \pm 1}{2}$$

$$= \begin{cases} -4 & \text{by taking the positive square root} \\ -5 & \text{by taking the negative square root} \end{cases}$$

The two solutions are therefore $x = -4$ and $x = -5$.

12.15 **Solve the equation $2x^2 - 3x - 7 = 0$ using the formula.**

Solution In this example $a = 2$, $b = -3$ and $c = -7$. Care should be taken with the negative signs. Substituting these into the formula we find

$$x = \frac{-(-3) \pm \sqrt{(-3)^2 - 4(2)(-7)}}{2(2)}$$

$$= \frac{3 \pm \sqrt{9 + 56}}{4}$$

$$= \frac{3 \pm \sqrt{65}}{4}$$

$$= \frac{3 \pm 8.062}{4}$$

$$= \begin{cases} 2.766 & \text{by taking the positive square root} \\ -1.266 & \text{by taking the negative square root} \end{cases}$$

The two solutions are therefore $x = 2.766$ and $x = -1.266$.

If the values of a, b and c are such that $b^2 - 4ac$ is positive the formula will produce two solutions known as **distinct real roots** of the equation. If $b^2 - 4ac = 0$ there will be a single root known as a **repeated root**. Some books refer to the equation having **equal roots**. If the equation is such that $b^2 - 4ac$ is negative the formula requires us to find the square root of a negative number. In ordinary arithmetic this is impossible and we say that in such a case the quadratic equation does not possess any real roots.

The quantity $b^2 - 4ac$ is called the **discriminant**, because it allows us to distinguish between the three possible cases. In summary:

Key point

Given

$$ax^2 + bx + c = 0$$

then

$b^2 - 4ac > 0$ two distinct real roots

$b^2 - 4ac = 0$ repeated (equal) root

$b^2 - 4ac < 0$ no real roots

WORKED EXAMPLES

12.16 Can the equation $x^2 + 49 = 0$ be solved using the formula?

Solution In this example $a = 1$, $b = 0$ and $c = 49$. Therefore $b^2 - 4ac = 0 - 4(1)(49)$, which is negative. Therefore this equation does not possess any real roots.

12.17 Use the formula to solve the equation $4x^2 + 4x + 1 = 0$.

Solution In this example $a = 4$, $b = 4$ and $c = 1$. Applying the formula gives

$$x = \frac{-4 \pm \sqrt{4^2 - 4(4)(1)}}{(2)(4)}$$

$$= \frac{-4 \pm \sqrt{16 - 16}}{8}$$

$$= \frac{-4 \pm 0}{8} = -\frac{1}{2}$$

There is a single, repeated root $x = -\frac{1}{2}$. Note that $b^2 - 4ac = 0$.

Self-assessment questions 12.2

1. Under what conditions will a quadratic equation possess distinct real roots?
2. Under what conditions will a quadratic equation possess a repeated root?

Exercise 12.3

1. Solve the following quadratic equations by factorisation:
   (a) $x^2 + x - 2 = 0$ (b) $x^2 - 8x + 15 = 0$ (c) $4x^2 + 6x + 2 = 0$
   (d) $x^2 - 6x + 9 = 0$ (e) $x^2 - 81 = 0$ (f) $x^2 + 4x + 3 = 0$
   (g) $x^2 + 2x - 3 = 0$ (h) $x^2 + 3x - 4 = 0$ (i) $x^2 + 6x + 5 = 0$
   (j) $x^2 - 12x + 35 = 0$ (k) $x^2 + 12x + 35 = 0$ (l) $2x^2 + x - 3 = 0$
   (m) $2x^2 - x - 6 = 0$ (n) $2x^2 - 7x - 15 = 0$ (o) $3x^2 - 2x - 1 = 0$
   (p) $9x^2 - 12x - 5 = 0$ (q) $7x^2 + x = 0$ (r) $4x^2 + 12x + 9 = 0$

2. Solve the following quadratic equations using the formula:
   (a) $3x^2 - 6x - 5 = 0$ (b) $x^2 + 3x - 77 = 0$ (c) $2x^2 - 9x + 2 = 0$
   (d) $x^2 + 3x - 4 = 0$ (e) $3x^2 - 3x - 4 = 0$ (f) $4x^2 + x - 1 = 0$
   (g) $x^2 - 7x - 3 = 0$ (h) $x^2 + 7x - 3 = 0$ (i) $11x^2 + x + 1 = 0$
   (j) $2x^2 - 3x - 7 = 0$

3. Solve the following quadratic equations:
   (a) $6x^2 + 13x + 6 = 0$ (b) $3t^2 + 13t + 12 = 0$ (c) $t^2 - 7t + 3 = 0$

Test and assignment exercises 12.1

1. Solve the following equations:
   (a) $13t - 7 = 2t + 5$ (b) $-3t + 9 = 13 - 7t$ (c) $-5t = 0$

2. Solve the following equations:
   (a) $4x = 16$ (b) $\frac{x}{12} = 9$ (c) $4x - 13 = 3$ (d) $4x - 14 = 2x + 8$
   (e) $3x - 17 = 4$ (f) $7 - x = 9 + 3x$ (g) $4(2 - x) = 8$

3. Solve the following equations:
   (a) $2y = 8$ (b) $5u = 14u + 3$ (c) $5 = 4I$ (d) $13i + 7 = 2i - 9$
   (e) $\frac{1}{4}x + 9 = 3 - \frac{1}{2}x$ (f) $\frac{4x + 2}{2} + 8x = 0$

4. Solve the following equations:
   (a) $3x^2 - 27 = 0$ (b) $2x^2 + 18x + 14 = 0$ (c) $x^2 = 16$
   (d) $x^2 - x - 72 = 0$ (e) $2x^2 - 3x - 44 = 0$ (f) $x^2 - 4x - 21 = 0$

5. The solutions of the equation $x^2 + 4x = 0$ are identical to the solutions of $3x^2 + 12x = 0$. True or false?

6. Solve the following simultaneous equations:
   (a) $x - 2y = -11,\quad 7x + y = -32$ (b) $2x - y = 2,\quad x + 3y = 29$
   (c) $x + y = 19,\quad -x + y = 1$

The straight line

Objectives: This chapter:

- describes some special properties of straight line graphs
- explains the equation $y = mx + c$
- explains the terms 'vertical intercept' and 'gradient'
- shows how the equation of a line can be calculated
- explains what is meant by a tangent to a curve
- explains what is meant by the gradient of a curve
- explains how the gradient of a curve can be estimated by drawing a tangent

Straight line graphs

Any equation of the form $y = mx + c$, where m and c are constants, will have a straight line graph. For example,

$$y = 3x + 7 \qquad y = -2x + \frac{1}{2} \qquad y = 3x - 1.5$$

will all result in straight line graphs. It is important to note that the variable x only occurs to the power 1. The values of m and c may be zero, so that $y = -3x$ is a straight line for which the value of c is zero, and $y = 17$ is a straight line for which the value of m is zero.

Key point

Any straight line has an equation of the form $y = mx + c$ where m and c are constants.

WORKED EXAMPLES

12.18 **Which of the following equations have straight line graphs? For those that do, identify the values of m and c.**

(a) $y=7x+5$ (b) $y=\dfrac{13x-5}{2}$ (c) $y=3x^2+x$ (d) $y=-19$

Solution

(a) $y=7x+5$ is an equation of the form $y=mx+c$ where $m=7$ and $c=5$. This equation has a straight line graph.

(b) The equation

$$y=\frac{13x-5}{2}$$

can be written as $y=\frac{13}{2}x-\frac{5}{2}$, which is in the form $y=mx+c$ with $m=\frac{13}{2}$ and $c=-\frac{5}{2}$. This has a straight line graph.

(c) $y=3x^2+x$ contains the term x^2. Such a term is not allowed in the equation of a straight line. The graph will not be a straight line.

(d) $y=-19$ is in the form of the equation of a straight line for which $m=0$ and $c=-19$.

12.19 **Plot each of the following graphs: $y=2x+3$, $y=2x+1$ and $y=2x-2$. Comment upon the resulting lines.**

Solution

The three graphs are shown in Figure 12.1. Note that all three graphs have the same steepness or **slope**. However, each one cuts the vertical

Figure 12.1
Graphs of $y=2x+3$, $y=2x+1$ and $y=2x-2$

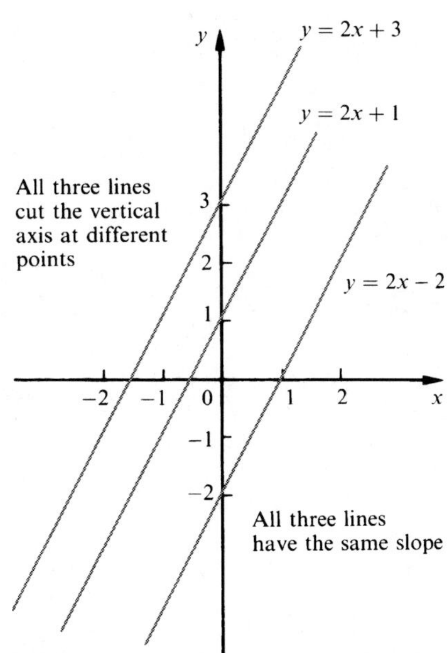

axis at a different point. This point can be obtained directly from the equation $y = 2x + c$ by looking at the value of c. For example, $y = 2x + 1$ intersects the vertical axis at $y = 1$. The graph of $y = 2x + 3$ cuts the vertical axis at $y = 3$, and the graph of $y = 2x - 2$ cuts this axis at $y = -2$.

The point where a graph cuts the vertical axis is called the **vertical intercept**. The vertical intercept can be obtained from $y = mx + c$ by looking at the value of c.

Key point

In the equation $y = mx + c$ the value of c gives the y coordinate of the point where the line cuts the vertical axis.

WORKED EXAMPLE

12.20 On the same graph plot the following: $y = x + 2$, $y = 2x + 2$, $y = 3x + 2$. Comment upon the graphs.

Solution The three straight lines are shown in Figure 12.2. Note that the steepness of the line is determined by the coefficient of x, with $y = 3x + 2$ being steeper than $y = 2x + 2$, and this in turn being steeper than $y = x + 2$.

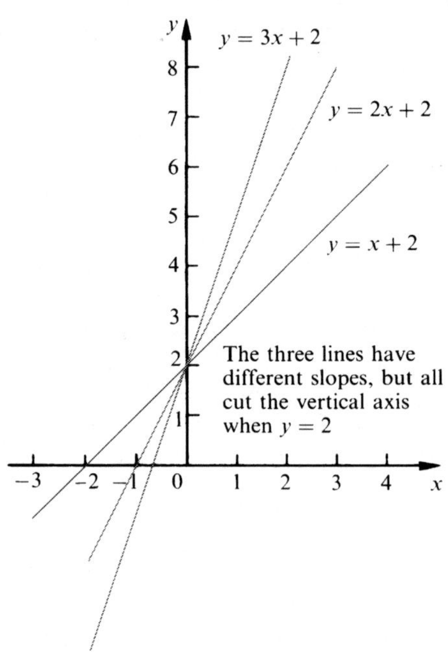

Figure 12.2 Graphs of $y = x + 2$, $y = 2x + 2$, and $y = 3x + 2$

The value of m in the equation $y = mx + c$ determines the steepness of the straight line. The larger the value of m, the steeper is the line. The value m is known as the slope or **gradient** of the straight line.

Key point

In the equation $y = mx + c$ the value m is known as the gradient and is a measure of the steepness of the line.

If m is positive, the line will rise as we move from left to right. If m is negative, the line will fall. If $m = 0$ the line will be horizontal. We say it has zero gradient. These points are summarised in Figure 12.3.

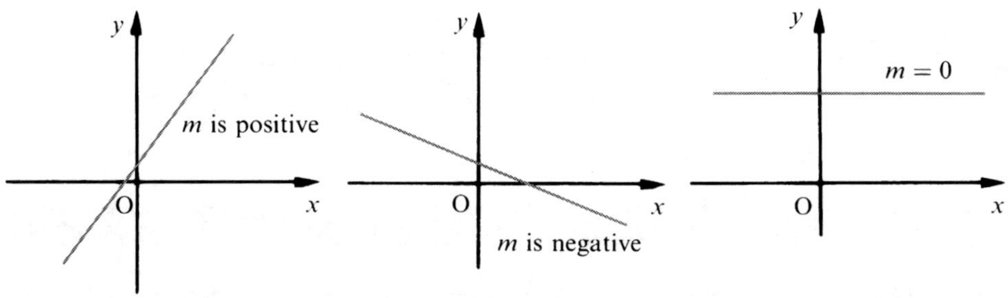

Figure 12.3
Straight line graphs with positive, negative and zero gradients

Self-assessment questions 12.3

1. Give the standard form of the equation of a straight line.
2. Explain the meaning of m and c in the equation $y = mx + c$.

Exercise 12.4

1. Identify, without plotting, which of the following functions will give straight line graphs:
   (a) $y = 3x + 9$ (b) $y = -9x + 2$ (c) $y = -6x$ (d) $y = x^2 + 3x$
   (e) $y = 17$ (f) $y = x^{-1} + 7$
2. Identify the gradient and vertical intercept of each of the following lines:
   (a) $y = 9x - 11$ (b) $y = 8x + 1.4$ (c) $y = \frac{1}{2}x - 11$ (d) $y = 17 - 2x$
   (e) $y = \dfrac{2x + 1}{3}$ (f) $y = \dfrac{4 - 2x}{5}$ (g) $y = 3(x - 1)$ (h) $y = 4$
3. Identify (i) the gradient and (ii) the vertical intercept of the following lines:
   (a) $y + x = 6$ (b) $y - 2x + 1 = 0$ (c) $2y - 4x + 3 = 0$ (d) $3x - 4y + 12 = 0$
   (e) $3x + \frac{y}{2} - 9 = 0$

Finding the equation of a straight line from its graph

If we are given the graph of a straight line it is often necessary to find its equation, $y = mx + c$. This amounts to finding the values of m and c. Finding the vertical intercept is straightforward because we can look directly for the point where the line cuts the y axis. The y coordinate of this point gives the value of c. The gradient m can be determined from knowledge of any two points on the line using the formula

Key point

$$\text{gradient} = \frac{\text{difference between the } y \text{ coordinates}}{\text{difference between the } x \text{ coordinates}}$$

WORKED EXAMPLES

12.21 A straight line graph is shown in Figure 12.4. Determine its equation.

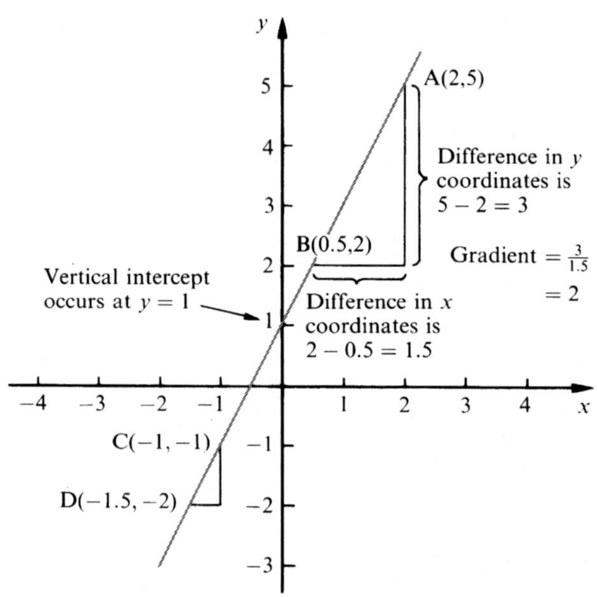

Figure 12.4 Graph for Example 12.21

Solution We require the equation of the line in the form $y = mx + c$. From the graph it is easy to see that the vertical intercept occurs at $y = 1$. Therefore the value of c is 1. To find the gradient m we choose any two points on the line. We have chosen the point A with coordinates (2,5) and the point B with coordinates (0.5, 2). The difference between their y coordinates is then

$5-2=3$. The difference between their x coordinates is $2-0.5=1.5$. Then

$$\text{gradient} = \frac{\text{difference between their } y \text{ coordinates}}{\text{difference between their } x \text{ coordinates}}$$

$$= \frac{3}{1.5} = 2$$

The gradient m is equal to 2. Note that as we move from left to right the line is rising and so the value of m is positive. The equation of the line is then $y=2x+1$. There is nothing special about the points A and B. Any two points are sufficient to find m. For example, using the points C with coordinates $(-1, -1)$ and D with coordinates $(-1.5, -2)$ we would find

$$\text{gradient} = \frac{\text{difference between their } y \text{ coordinates}}{\text{difference between their } x \text{ coordinates}}$$

$$= \frac{-1-(-2)}{-1-(-1.5)}$$

$$= \frac{1}{0.5} = 2$$

as before.

12.22 **A straight line graph is shown in Figure 12.5. Find its equation.**

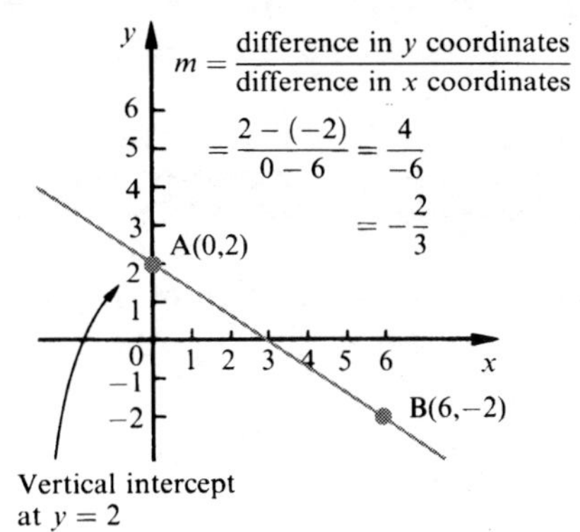

Figure 12.5 Graph for Worked Example 12.22

Solution We need to find the equation in the form $y=mx+c$. From the graph we see immediately that the value of c is 2. To find the gradient we have selected any two points, $A(0, 2)$ and $B(6, -2)$. The difference between their y coordinates is $2-(-2)=4$. The difference between their x coordinates is $0 - 6 = -6$. Then

$$\text{gradient} = \frac{\text{difference between their } y \text{ coordinates}}{\text{difference between their } x \text{ coordinates}}$$

$$= \frac{4}{-6}$$

$$= -\frac{2}{3}$$

The equation of the line is therefore $y = -\frac{2}{3}x + 2$. Note in particular that, because the line is sloping downwards as we move from left to right, the gradient is negative. Note also that the coordinates of A and B both satisfy the equation of the line. That is, for $A(0, 2)$,

$$2 = -\frac{2}{3}(0) + 2$$

and for $B(6, -2)$,

$$-2 = -\frac{2}{3}(6) + 2$$

The coordinates of any other point on the line must also satisfy the equation.

The point noted at the end of Worked Example 12.5 is important:

Key point

If the point (a, b) lies on the line $y = mx + c$ then this equation is satisfied by letting $x = a$ and $y = b$.

WORKED EXAMPLE

12.23 Find the equation of the line shown in Figure 12.6.

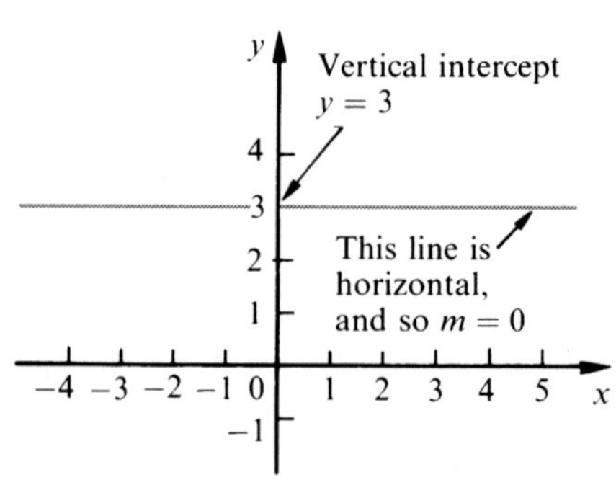

Figure 12.6 Graph for Example 12.23

Solution We are required to express the equation in the form $y = mx + c$. From the graph we notice that the line is horizontal. This means that its gradient is 0, that is, $m = 0$. Furthermore the line cuts the vertical axis at $y = 3$ and so the equation of the line is $y = 0x + 3$ or simply $y = 3$.

It is not necessary to sketch a graph in order to find the equation. Consider the following worked examples, which illustrate an algebraic method.

WORKED EXAMPLES

12.24 A straight line passes through $A(7, 1)$ and $B(-3, 2)$. Find its equation.

Solution The equation must be of the form $y = mx + c$. The gradient of the line can be found from

$$\text{gradient} = m = \frac{\text{difference between their } y \text{ coordinates}}{\text{difference between their } x \text{ coordinates}}$$

$$= \frac{1 - 2}{7 - (-3)}$$

$$= \frac{-1}{10}$$

$$= -0.1$$

Hence $y = -0.1x + c$. We can find c by noting that the line passes through (7,1), that is, the point where $x = 7$ and $y = 1$. Substituting these values into the equation $y = -0.1x + c$ gives

$$1 = -0.1(7) + c$$

so that $c = 1 + 0.7 = 1.7$. Therefore the equation of the line is $y = -0.1x + 1.7$.

12.25 Determine the equation of the line that passes through $(4, -1)$ and has gradient -2.

Solution Let the equation of the line be $y = mx + c$. We are told that the gradient of the line is -2, that is, $m = -2$ and so we have

$$y = -2x + c$$

The point $(4, -1)$ lies on this line: hence when $x = 4$, $y = -1$. These values are substituted into the equation of the line:

$$-1 = -2(4) + c$$

$$c = 7$$

The equation of the line is thus $y = -2x + 7$.

Self-assessment questions 12.4

1. State the formula for finding the gradient of a straight line when two points upon it are known. If the two points are (x_1, y_1) and (x_2, y_2) write down an expression for the gradient.
2. Explain how the value of c in the equation $y = mx + c$ can be found by inspecting the straight line graph.

Exercise 12.5

1. A straight line passes through the two points (1,7) and (2,9). Sketch a graph of the line and find its equation.
2. Find the equation of the line that passes through the two points (2,2) and (3,8).
3. Find the equation of the line that passes through (8,2) and (−2,2).
4. Find the equation of the straight line that has gradient 1 and passes through the origin.
5. Find the equation of the straight line that has gradient −1 and passes through the origin.
6. Find the equation of the straight line passing through (−1, 6) with gradient 2.
7. Which of the following points lie on the line $y = 4x - 3$?
   (a) (1,2) (b) (2,5) (c) (5, 17)
   (d) (−1,−7) (e) (0,2)
8. Find the equation of the straight line passing through (−3, 7) with gradient −1.
9. Determine the equation of the line passing through (−1, −6) that is parallel to the line $y = 3x + 17$.
10. Find the equation of the line with vertical intercept −2 passing through (3,10).

Gradients of tangents to curves

Figure 12.7 shows a graph of $y = x^2$. If you study the graph you will notice that as we move from left to right, at some points the y values are decreasing, whereas at others the y values are increasing. It is intuitively obvious that the slope of the curve changes from point to point. At some points, such as A, the curve appears quite steep and falling. At points such as B the curve appears quite steep and rising. Unlike a straight line, the slope of a curve is not fixed but changes as we move from one point to another. A useful way of measuring the slope at any point is to draw a **tangent** to the curve at that point. The tangent is a straight line that just touches the curve at the point of interest. In Figure 12.7 a tangent to the curve $y = x^2$ has been drawn at the point (2, 4). If we calculate the

Figure 12.7
A graph of $y = x^2$

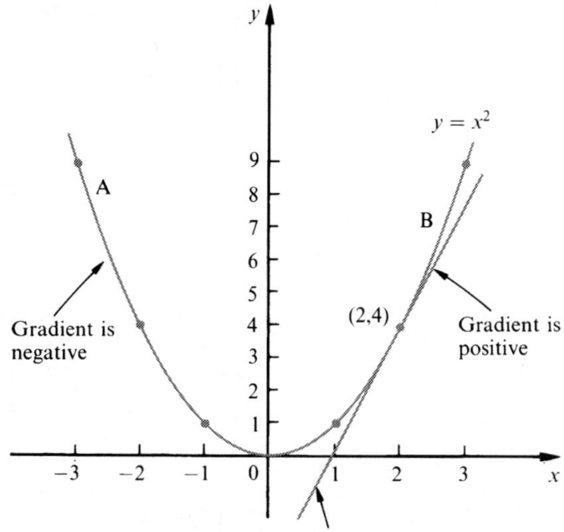

gradient of this tangent, this gives the gradient of the curve at the point (2, 4).

Key point

The gradient of a curve at any point is equal to the gradient of the tangent at that point.

WORKED EXAMPLE

12.26

(a) Plot a graph of $y = x^2 - x$ for values of x between -2 and 4.

(b) Draw in tangents at the points $A(-1, 2)$ and $B(3, 6)$.

(c) By calculating the gradients of these tangents find the gradient of the curve at A and at B.

Solution

(a) A table of values and the graph is shown in Figure 12.8.

(b) We now draw tangents at A and B. At present, the best we can do is estimate these by eye.

(c) We now calculate the gradient of the tangent at A. We select any two points on the tangent and calculate the difference between their y coordinates and the difference between their x coordinates. We have chosen the points $(-3, 8)$ and $(-2, 5)$. Referring to Figure 12.8 we see that

$$\text{gradient of tangent at } A = \frac{8-5}{-3-(-2)} = \frac{3}{-1}$$

$$= -3$$

Figure 12.8
A graph of $y = x^2 - x$

$y = x^2 - x$

x	-2	-1	0	1	2	3	4
y	6	2	0	0	2	6	12

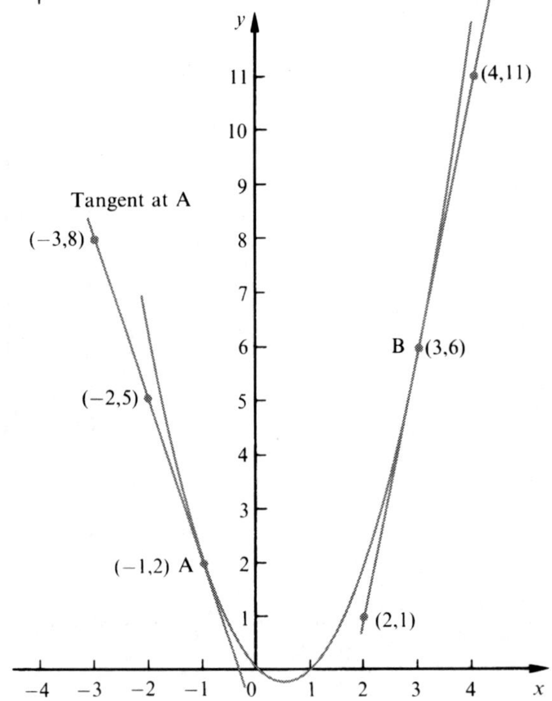

Hence the gradient of the curve at A is -3. Similarly, to find the gradient of the tangent at B we have selected two points on this tangent, namely (4, 11) and (2, 1). We find

$$\text{gradient of tangent at } B = \frac{11-1}{4-2} = \frac{10}{2}$$

$$= 5$$

Hence the gradient of the tangent at B is 5. Thus the gradient of the curve at B(3, 6) is 5.

Clearly, the accuracy of our answer depends to a great extent upon how well we can draw and measure the gradient of the tangent.

WORKED EXAMPLE

12.27 (a) Sketch a graph of the curve $y = x^3$ for $-2 \leqslant x \leqslant 2$.

(b) Draw the tangent to the graph at the point where $x = 1$.
(c) Estimate the gradient of this tangent and find its equation.

Solution (a) A graph is shown in Figure 12.9.

Figure 12.9
Graph of $y = x^3$

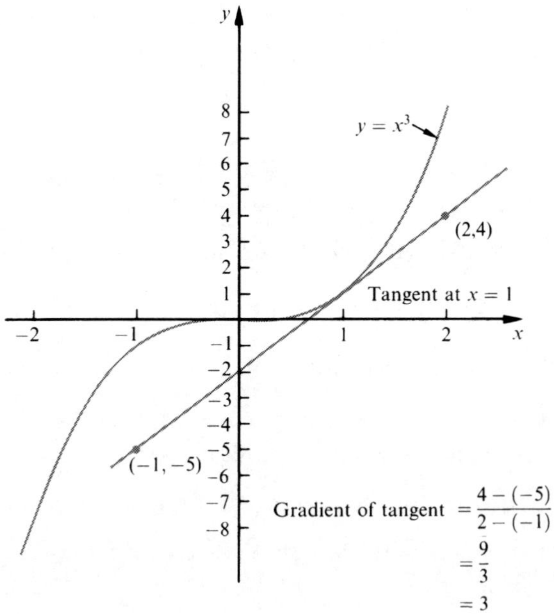

(b) The tangent has been drawn at $x = 1$.
(c) Let us write the equation of the tangent as $y = mx + c$. Two points on the tangent have been selected in order to estimate the gradient. These are (2, 4) and (−1, −5). From these we find

$$\text{gradient of tangent is approximately } \frac{4 - (-5)}{2 - (-1)} = \frac{9}{3} = 3$$

Therefore $m = 3$. The value of c is found by noting that the vertical intercept of the tangent is −2. The equation of the tangent is then $y = 3x - 2$.

Of course, this method will usually result in an approximation based upon how well we have drawn the graph and its tangent.

Self-assessment questions 12.5

1. Explain what is meant by the 'tangent' to a curve at a point.
2. Explain how a tangent is used to determine the gradient of a curve.

Exercise 12.6

1. Draw the graph of $y = 2x^2 - 1$ for values of x between -3 and 3. By drawing a tangent, estimate the gradient of the curve at $A(2, 7)$ and $B(-1, 1)$.
2. Draw the graph of $y = -2x^2 + 2$ for values of x between -3 and 3. Draw the tangent at the point where $x = 1$ and calculate its equation.

Test and assignment exercises 12.2

1. Which of the following will have straight line graphs?

   (a) $y = 2x - 11$ (b) $y = 5x + 10$ (c) $y = x^2 - 1$ (d) $y = -3 + 3x$ (e) $y = \dfrac{2x+3}{2}$

   For each straight line, identify the gradient and vertical intercept.
2. Find the equation of the straight line that passes through the points (1, 11) and (2, 18). Show that the line also passes through $(-1, -3)$.
3. Find the equation of the line that has gradient -2 and passes through the point (1, 1).
4. Find the equation of the line that passes through $(-1, 5)$ and (1, 5). Does the line also pass through (2, 6)?
5. Draw a graph of $y = -x^2 + 3x$ for values of x between -3 and 3. By drawing in tangents estimate the gradient of the curve at the points $(-2, -10)$ and (1, 2).
6. Find the equations of the lines passing through the origin with gradients (a) -2, (b) -4, (c) 4.
7. Find the equation of the line passing through (4,10) and parallel to $y = 6x - 3$.
8. Find the equation of the line with vertical intercept 3 and passing through $(-1, 9)$.
9. Find where the line joining $(-2, 4)$ and (3, 10) cuts (a) the x axis (b) the y axis.
10. A line cuts the x axis at $x = -2$ and the y axis at $y = 3$. Determine the equation of the line.
11. Determine where the line $y = 4x - 1$ cuts

    (a) the y axis (b) the x axis (c) the line $y = 2$.

Straight line graphs in biology

Straight line graphs are extremely important and useful to biologists. Key aspects of a straight line graph are the gradient and the intercepts on the x- and y-axes. Straight line graphs are described by the equation $y = mx + c$ where x is the gradient and c is the intercept on the y axis. Many biological problems use straight line graphs to find specific values needed to solve equations and answers to specific questions. For straight line graphs, when the intercept $c = 0$, the value on the x axis is directly proportional to the value on the y axis.

Figure 12.9a

Graphs of different gradients

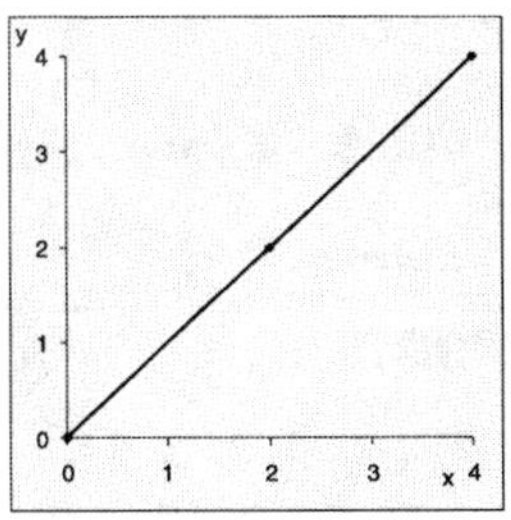

In this case, $y = x$, so when $y = 2$, $x = 2$, when $y = 4$, $x = 4$.

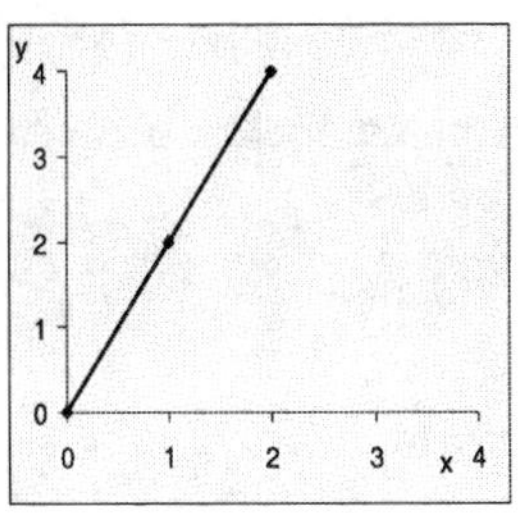

When $y = 2$, $x = 1$
When $y = 4$, $x = 2$

y is equal to twice x, or $y = 2x$

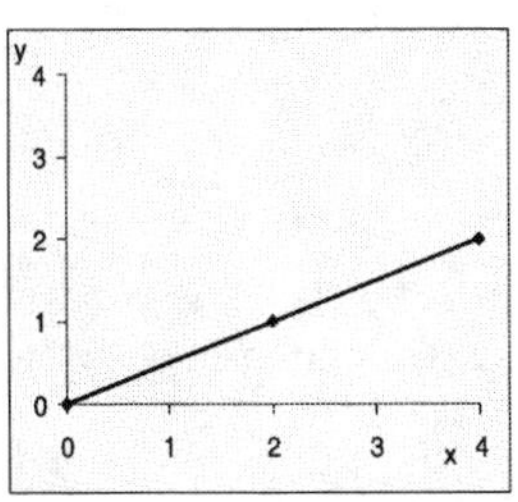

When $y = 1$, $x = 2$
When $y = 2$, $x = 4$

So $2y = x$ or $y = \frac{1}{2}x$

So the number before x tells you the slope of the line. In other words, it tells you about the **gradient** of the line. The gradient is assigned the letter m.

For example:
If $y = 2x$, $m=2$
If $y = ½\,x$, $m = ½$

$y = mx$, where m is the gradient of the line, or
m = the rate of change of y with respect to x.

Examples of straight-line relationships

Figure 12.9b Ants consumed against time

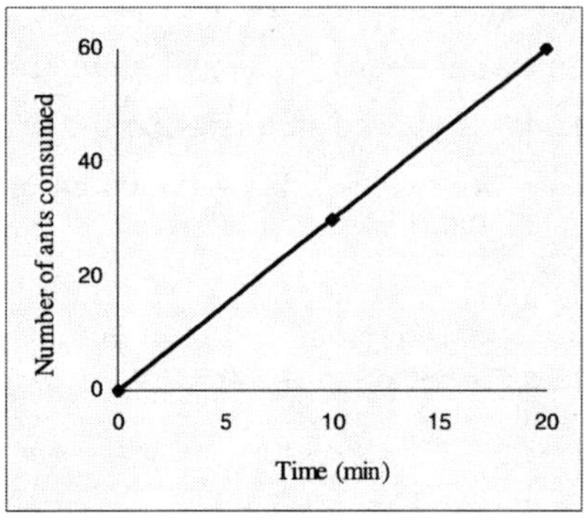

$y = 3x$

y = number of ants consumed

x = time

m = rate at which the number of ants are consumed over time

$$m = \frac{change\ in\ y}{change\ in\ x} = \frac{\Delta y}{\Delta x}$$

Taking the highest values for x and y from the graph:

$$m = \frac{60\ ants\ consumed}{20\ min} = 3\ ants\ consumed\ \text{min}^{-1}$$

Figure 12.9c Plants counted against time

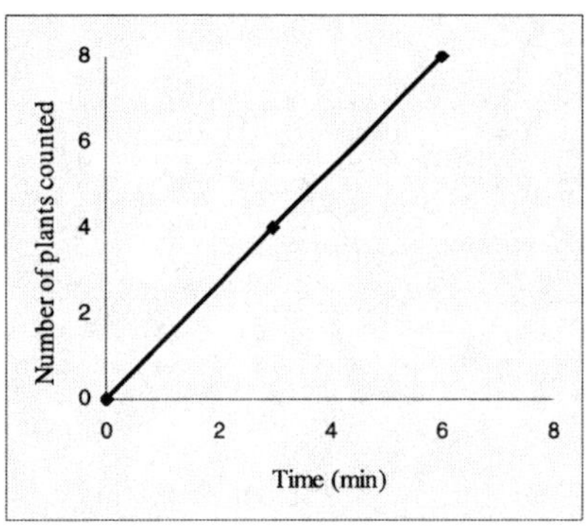

$y = 1.33x$

y = number of plants counted

x = time

m = rate at which the number of plants are counted over time

$$m = \frac{\Delta y}{\Delta x}$$ As above,

$$m = \frac{8\ plants\ counted}{6\ min} = 1.33\,plants\ \text{min}^{-1}$$

This tells you that the biologist is counting plants at a **rate** of 1.33 plants per minute.

Figure 12.9d Rate of consumption of bacteria against temperature

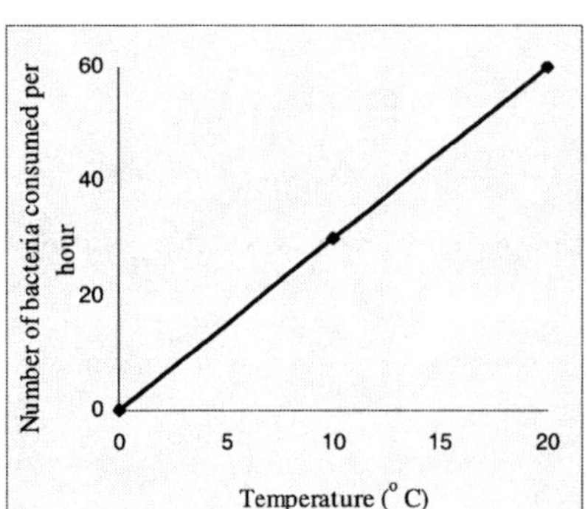

$y = 3x$

y = number of bacteria consumed per hour

x = temperature (°C)

m = rate at which the number of bacteria are consumed as temperature increases

$$m = \frac{\Delta y}{\Delta x}$$ As above,

$$m = \frac{number\ bacteria\ consumed\ h^{-1}}{temperature\ {}^{o}C} = \frac{60\ bacteria\ h^{-1}}{20\ {}^{o}C}$$

As the temperature increases, for each °C three more bacteria are consumed per hour, or m = 3 bacteria $h^{-1}\ {}^{o}C^{-1}$

While the majority of straight line graphs you will come across will be in the positive section of a graph (where the numbers on the x and y-axes are all positive) it is important to understand that graphs have four quadrants, where values on the x or y axes can be positive or negative. Also many data sets do not intercept the x and y-axes at zero. A graph with an intercept on the y axis that is positive looks like this:

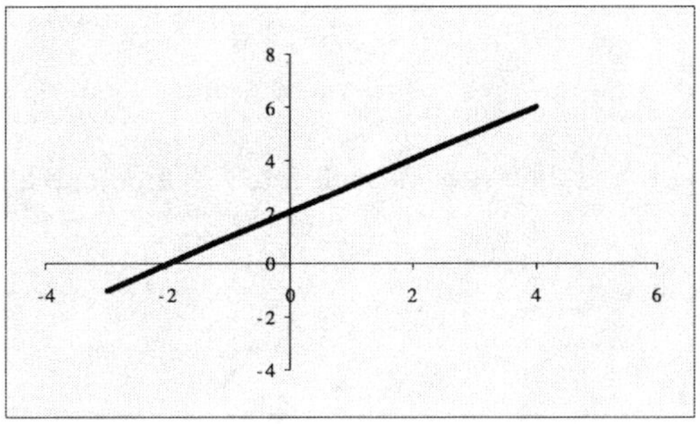

Figure 12.9e
Graph with a positive y intercept and a positive gradient

$$y = mx + c$$

And like this if c is a negative intercept:

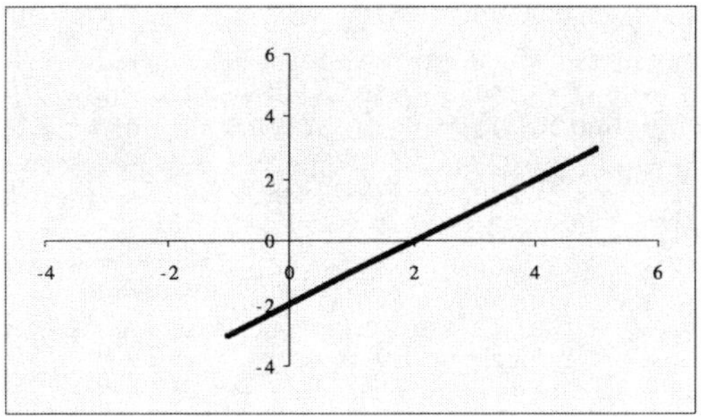

Figure 12.9f
Graph with a negative y intercept and a positive gradient

$$y = mx - c$$

If the graph has a negative slope it will look like this:

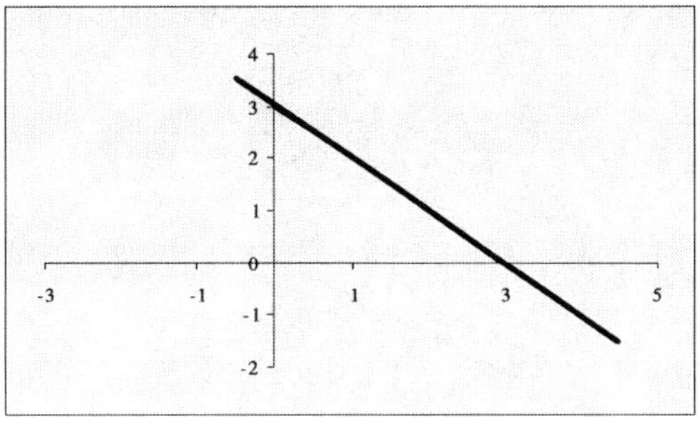

Figure 12.9g
Graph with a positive y intercept and a negative gradient

$$y = -mx + c$$

So the $y = mx + c$ equation tells you all you need to know about the position and slope of a straight line graph.

Test your understanding.

1. If the growth rate of 2 cheetah cubs in their first week of life followed the equations:

a) $y = 2.4x + 3$

b) $y = 3.5x + 2$

where y = mass and x = time. Which cub had the faster growth rate? Which cub had the higher birth mass?

2. Sketch the following equations as graphs:

a) $y = 0.5x - 2$

b) $y = -2x + 1$

c) $y = 3x - 1$

3. The Beer Lambert Law uses the equation $A = Ecl$ where:

A is the absorbance, and has no units

E = extinction coefficient ($l\ mol^{-1}\ cm^{-1}$)

l = path length (cm), usually 1 cm

c= concentration ($mol\ l^{-1}$)

a) If E is the gradient, what is plotted on the x- and y-axes?

b) If the graph followed the equation $y = 3.24x$ what is the value of E, and where does the line of the graph intercept the y axis?

4. From the following graph, which shows how the distance travelled by a tortoise changes with time, what is the speed of the tortoise?

Figure 12.9h
Speed of a tortoise

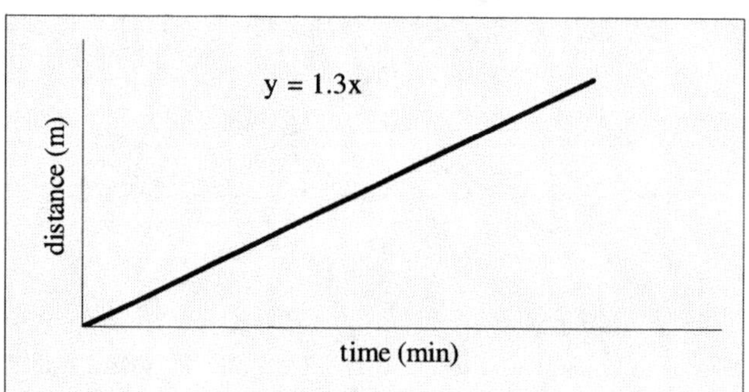

03

The exponential function

Objectives: This chapter:

- shows how to simplify exponential expressions
- describes the form of the exponential function
- illustrates graphs of exponential functions
- lists the properties of the exponential function
- shows how to solve equations with exponential terms using a graphical technique

Exponential expressions

An expression of the form a^x is called an **exponential expression**. The number a is called the **base** and x is the **power** or **index**. For example, 4^x and 0.5^x are both exponential expressions. The first has base 4 and the second has base 0.5. An exponential expression can be evaluated using the power button on your calculator. For example, the exponential expression with base 4 and index 0.3 is

The letter e always denotes the constant 2.71828....

$$4^{0.3} = 1.516$$

One of the most commonly used values for the base in an exponential expression is 2.71828.... This number is denoted by the letter e.

Key point

The most common exponential expression is

$$e^x$$

where e is the exponential constant, 2.71828....

The expression e^x is found to occur in the modelling of many natural phenomena, for example population growth, spread of bacteria and radioactive decay. Scientific calculators are pre-programmed to evaluate e^x for any value of x. Usually the button is marked e^x or $\exp x$. Use the next worked example to check you can use this facility on your calculator.

WORKED EXAMPLE

12.28 Use a scientific calculator to evaluate

(a) e^2 (b) e^3 (c) $e^{1.3}$ (d) $e^{-1.3}$

Solution We use the e^x button giving:

(a) $e^2 = 7.3891$

(b) $e^3 = 20.0855$

(c) $e^{1.3} = 3.6693$

(d) $e^{-1.3} = 0.2725$

Exponential expressions can be simplified using the normal rules of algebra. The laws of indices apply to exponential expressions. We note

$$e^a e^b = e^{a+b}$$

$$\frac{e^a}{e^b} = e^{a-b}$$

$$e^0 = 1$$

$$(e^a)^b = e^{ab}$$

WORKED EXAMPLES

12.29 Simplify the following exponential expressions:

(a) $e^2 e^4$ (b) $\dfrac{e^4}{e^3}$ (c) $(e^2)^{2.5}$

Solution (a) $e^2 e^4 = e^{2+4} = e^6$

(b) $\dfrac{e^4}{e^3} = e^{4-3} = e^1 = e$

(c) $(e^2)^{2.5} = e^{2 \times 2.5} = e^5$

12.30 Simplify the following exponential expressions:

(a) $e^x e^{3x}$ (b) $\dfrac{e^{4t}}{e^{3t}}$ (c) $(e^t)^4$

Solution

(a) $e^x e^{3x} = e^{x+3x} = e^{4x}$

(b) $\dfrac{e^{4t}}{e^{3t}} = e^{4t-3t} = e^t$

(c) $(e^t)^4 = e^{4t}$

12.31 Simplify

(a) $\sqrt{e^{6t}}$ (b) $e^{3y}(1+e^y) - e^{4y}$ (c) $(2e^t)^2(3e^{-t})$

Solution

(a) $\sqrt{e^{6t}} = (e^{6t})^{1/2} = e^{6t/2} = e^{3t}$

(b) $e^{3y}(1+e^y) - e^{4y} = e^{3y} + e^{3y}e^y - e^{4y}$

$= e^{3y} + e^{4y} - e^{4y}$

$= e^{3y}$

$\sqrt{a} = a^{\frac{1}{2}}$, that is, a square root sign is equivalent to the power $\frac{1}{2}$.

(c) $(2e^t)^2(3e^{-t}) = (2e^t)(2e^t)(3e^{-t})$

$= 2.2.3.e^t e^t e^{-t}$

$= 12e^{t+t-t}$

$= 12e^t$

12.32 (a) Verify that

$$(e^t+1)^2 = e^{2t} + 2e^t + 1$$

(b) Hence simplify $\sqrt{e^{2t} + 2e^t + 1}$

Solution

(a) $(e^t+1)^2 = (e^t+1)(e^t+1)$

$= e^t e^t + e^t + e^t + 1$

$= e^{2t} + 2e^t + 1$

(b) $\sqrt{e^{2t} + 2e^t + 1} = \sqrt{(e^t+1)^2} = e^t + 1$

Exercise 12.7

1. Use a scientific calculator to evaluate

   (a) $e^{2.3}$ (b) $e^{1.9}$ (c) $e^{-0.6}$

   (d) $\dfrac{1}{e^{-2}+1}$ (e) $\dfrac{3e^2}{e^2-10}$

   (f) $e^2\left(e + \dfrac{1}{e}\right) - e$ (g) $e^{1.5}e^{2.7}$

   (h) $\sqrt{e}\sqrt{4e}$

2. Simplify

   (a) $e^2.e^7$ (b) $\dfrac{e^7}{e^4}$ (c) $\dfrac{e^{-2}}{e^3}$

   (d) $\dfrac{e^3}{e^{-1}}$ (e) $\dfrac{(4e)^2}{(2e)^3}$

3. Simplify each of the following expressions as far as possible:

   (a) $e^x e^{-3x}$ (b) $e^{3x}e^{-x}$

   (c) $e^{-3x}e^{-x}$ (d) $(e^{-x})^2 e^{2x}$

   (e) $\dfrac{e^{3x}}{e^x}$ (f) $\dfrac{e^{-3x}}{e^{-x}}$

4. Simplify as far as possible

(a) $\dfrac{e^t}{e^3}$ (b) $\dfrac{2e^{3x}}{4e^x}$ (c) $\dfrac{(e^x)^3}{3e^x}$

(d) $e^x + 2e^{-x} + e^x$ (e) $\dfrac{e^x e^y e^z}{e^{x/2-y}}$

(f) $\dfrac{(e^{t/3})^6}{e^t + e^t}$ (g) $\dfrac{e^t}{e^{-t}}$ (h) $\dfrac{1+e^{-t}}{e^{-t}}$

(i) $(e^z e^{-z/2})^2$ (j) $\dfrac{e^{-3+t}}{2e^t}$

(k) $\left(\dfrac{e^{-1}}{e^{-x}}\right)^{-1}$

The exponential function and its graph

The **exponential function** has the form

$$y = e^x$$

The function $y = e^x$ is found to occur in the modelling of many natural phenomena. Table 12.1 gives values of x and e^x. Using Table 12.1, a graph of $y = e^x$ can be drawn. This is shown in Figure 12.10. From the table and graph we note the following important properties of $y = e^x$:

(a) The exponential function is never negative.

(b) When $x = 0$, the function value is 1.

(c) As x increases, then e^x increases. This is known as **exponential growth**.

Table 12.1

x	−3	−2	−1	0	1	2	3
e^x	0.0498	0.1353	0.3679	1	2.7183	7.3891	20.086

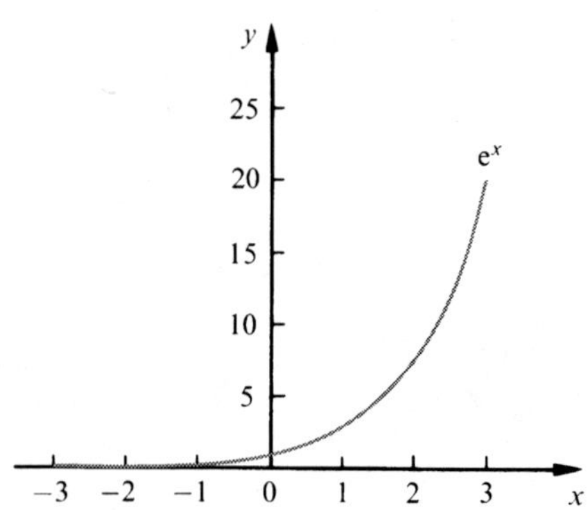

Figure 12.10
A graph of $y = e^x$

(d) By choosing x large enough, the value of e^x can be made larger than any given number. We say e^x increases without bound as x increases.

(e) As x becomes large and negative, e^x gets nearer and nearer to 0.

WORKED EXAMPLE

12.33 Sketch a graph of $y = 2e^x$ and $y = e^{2x}$ for $-3 \leqslant x \leqslant 3$.

Solution Table 12.2 gives values of $2e^x$ and e^{2x}, and the graphs are shown in Figure 12.11.

Table 12.2

x	-3	-2	-1	0	1	2	3
$2e^x$	0.0996	0.2707	0.7358	2	5.4366	14.7781	40.1711
e^{2x}	0.0025	0.0183	0.1353	1	7.3891	54.5982	403.429

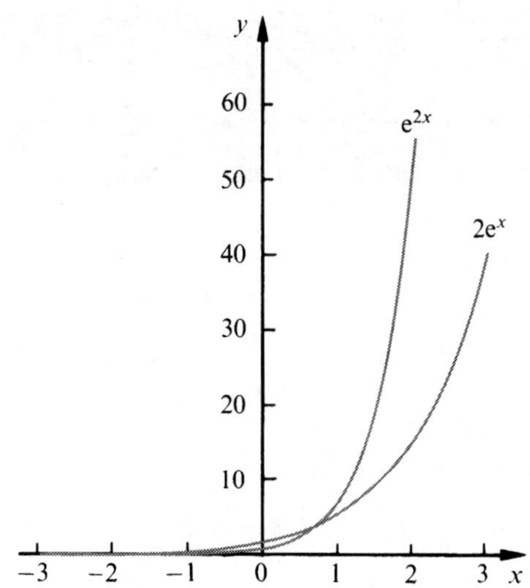

Figure 12.11
Graphs of $y = e^{2x}$ and $y = 2e^x$

Note that when $x > 0$, the graph of $y = e^{2x}$ rises more rapidly than that of $y = 2e^x$.

We now turn attention to an associated function: $y = e^{-x}$. Values are listed in Table 12.3 and the function is illustrated in Figure 12.12. From the table and the graph we note the following properties of $y = e^{-x}$:

(a) The function is never negative.

(b) When $x = 0$, the function has a value of 1.

(c) As x increases, then e^{-x} decreases, getting nearer to 0. This is known as **exponential decay**.

Table 12.3

x	−3	−2	−1	0	1	2	3
$-x$	3	2	1	0	−1	−2	−3
e^{-x}	20.086	7.3891	2.7183	1	0.3679	0.1353	0.0498

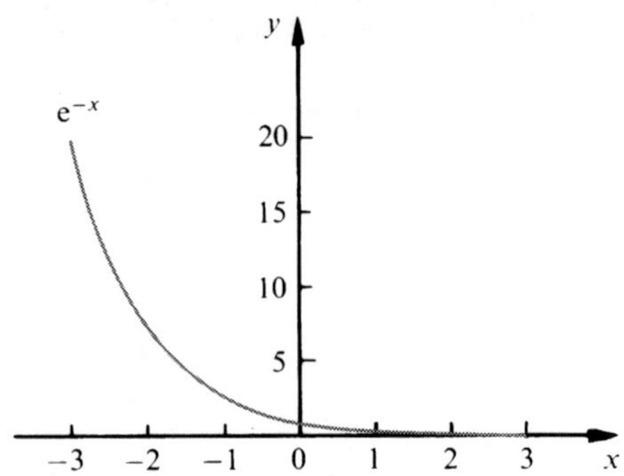

Figure 12.12
Graph of $y = e^{-x}$

Self-assessment questions 12.6

1. List properties that are common to $y = e^x$ and $y = e^{-x}$.
2. By choosing x large enough, then e^{-x} can be made to be negative. True or false?

Exercise 12.8

1. Draw up a table of values of $y = e^{3x}$ for x between −1 and 1 at intervals of 0.2. Sketch the graph of $y = e^{3x}$. Comment upon its shape and properties.
2. A species of animal has population $P(t)$ at time t given by

   $$P = 10 - 5e^{-t}$$

   (a) Sketch a graph of P against t for $0 \leqslant t \leqslant 5$.
   (b) What is the size of the population of the species as t becomes very large?
3. The concentration, $c(t)$, of a chemical in a reaction is modelled by the equation

   $$c(t) = 6 + 3e^{-2t}$$

   (a) Sketch a graph of c against t for $0 \leqslant t \leqslant 2$.
   (b) What value does the concentration approach as t becomes large?
4. The hyperbolic functions $\sinh x$ and $\cosh x$ are defined by

   $$\sinh x = \frac{e^x - e^{-x}}{2} \qquad \cosh x = \frac{e^x + e^{-x}}{2}$$

   (a) Show $e^x = \cosh x + \sinh x$.
   (b) Show $e^{-x} = \cosh x - \sinh x$.
5. Express $3 \sinh x + 7 \cosh x$ in terms of exponential functions.
6. Express $6e^x - 9e^{-x}$ in terms of $\sinh x$ and $\cosh x$.
7. Show $(\cosh x)^2 - (\sinh x)^2 = 1$.

Solving equations involving exponential terms using a graphical method

Many equations involving exponential terms can be solved using graphs. A graphical method will yield an approximate solution, which can then be refined by choosing a smaller interval for the domain. The following examples illustrate the technique.

WORKED EXAMPLE

12.34 (a) Plot $y = e^{x/2}$ and $y = 2e^{-x}$ for $-1 \leqslant x \leqslant 1$.

(b) Hence solve the equation

$$e^{x/2} - 2e^{-x} = 0$$

Solution (a) A table of values is drawn up for $e^{x/2}$ and $2e^{-x}$ so that the graphs can be drawn. Table 12.4 lists the values and Figure 12.13 shows the graphs of the functions.

Table 12.4

x	−1	−0.75	−0.5	−0.25	0	0.25	0.5	0.75	1
$e^{x/2}$	0.607	0.687	0.779	0.882	1	1.133	1.284	1.455	1.649
$2e^{-x}$	5.437	4.234	3.297	2.568	2	1.558	1.213	0.945	0.736

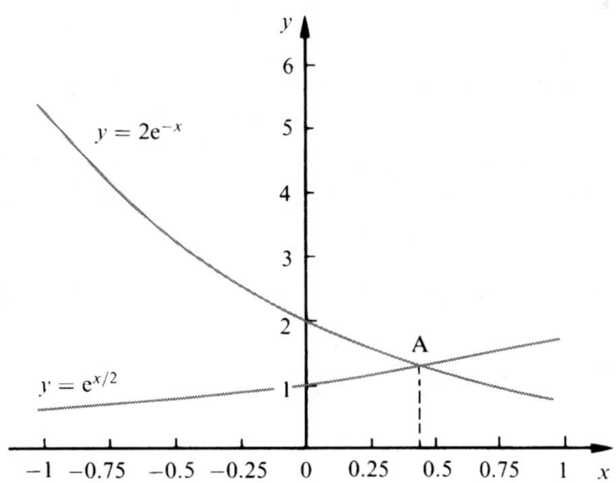

Figure 12.13
The graphs of $y = e^{x/2}$ and $y = 2e^{-x}$ intersect at A

(b) We note that the graphs intersect; the point of intersection is A. The equation

$$e^{x/2} - 2e^{-x} = 0$$

is equivalent to

$$e^{x/2} = 2e^{-x}$$

The graphs of $y = e^{x/2}$ and $y = 2e^{-x}$ intersect at A, where $x = 0.44$. Hence $x = 0.44$ is an approximate solution of $e^{x/2} - 2e^{-x} = 0$. The exact answer can be shown to be 0.46 (2 d.p.).

Exercise 12.9

1. (a) Plot $y = 15 - x^2$ and $y = e^x$ for $0 \leqslant x \leqslant 3$.
   (b) Hence find an approximate solution to
   $$e^x + x^2 = 15$$
2. (a) Plot $y = 12x^2 - 1$ and $y = 3e^{-x}$ for $-1 \leqslant x \leqslant 1$.
   (b) Hence state two approximate solutions of $3e^{-x} = 12x^2 - 1$.
3. By drawing $y = e^x$ for $-3 \leqslant x \leqslant 3$ and additional appropriate lines solve the following equations:
   (a) $e^x + x = 0$
   (b) $e^x - 1.5 = 0$
   (c) $e^x - x - 5 = 0$
   (d) $\frac{e^x}{2} + \frac{x}{2} - 5 = 0$
4. (a) Draw $y = 2 + 6e^{-t}$ for $0 \leqslant t \leqslant 3$
   (b) Use your graph to find an approximate solution to
   $$2 + 6e^{-t} = 5$$
   (c) Add the line $y = t + 4$ to your graph and hence find an approximate solution to
   $$6e^{-t} = t + 2$$

Test and assignment exercises 12.3

1. Use a calculator to evaluate
   (a) $e^{-3.1}$ (b) $e^{0.2}$ (c) $\frac{1}{e}$
2. (a) Sketch the graphs of
   $$y = e^{-x} \text{ and } y = \frac{x}{2}$$
   for $0 \leqslant x \leqslant 2$.
   (b) Hence solve the equation
   $$2e^{-x} - x = 0$$
3. Simplify where possible:
   (a) $e^x e^{2x} e^{-3x}$ (b) $\frac{e^{2x}e^{3x}}{e^{4x}}$ (c) $(e^{2x})^3 e^{-3x}$ (d) $e^{2x} + e^{3x} - e^{5x}$ (e) $\frac{e^{3x} + e^{5x}}{e^{2x}}$

4. Expand the brackets of:

(a) $(e^x - 1)^2$ (b) $(e^{-x} - 1)^2$ (c) $e^{2x}(e^x + 1)$ (d) $2e^x(e^{-x} + 3e^x)$

5. Simplify

(a) $e^t e^{t/2}$ (b) $e^2 e^3 e$ (c) $e^{3x}(e^x - e^{2x}) - (e^{2x})^2$ (d) $e^{3+t}e^{5-2t}$

(e) $\dfrac{e^{2z+3}e^{5-z}}{e^{z+2}}$ (f) $\sqrt{e^{4z}}$ (g) $(e^{x/2}e^x)^2$ (h) $\sqrt{e^{2t} + 2e^{2t+1} + e^{2t+2}}$

(i) $\dfrac{e^t + e^{2t}}{e^{2t} + e^{3t}}$

6. Use a graphical method to solve

(a) $e^{2x} = 2x + 2$ (b) $e^{-x} = 2 - x^2$ (c) $e^x = 2 + \dfrac{x^3}{2}$

7. (a) Sketch the graphs of

$y = 2 + e^{-x}$ and $y = e^x$

for $0 \leqslant x \leqslant 2$.

(b) Use your graphs to solve

$e^x - e^{-x} = 2$

8. Remove the brackets from the following expressions:

(a) $(e^{2t})^4$ (b) $e^t(e^t - e^{-t})$ (c) $e^{2y}(e^{3y} + e^{-2y} + e^y)$ (d) $(1 + e^t)^2$

9. The height of mercury in an experiment, $H(t)$, varies according to

$H(t) = 10 + 3e^{-t} - 4e^{-2t}$

What value does H approach as t gets very large?

10. Given

$$X(t) = \frac{3 + 2e^{2t}}{4 + e^{2t}}$$

find the value that X approaches as t gets very large.

Non-linear graphs in biology

Many relationships in biology do not follow straight lines. They do, however, often follow graphs that relate to specific equations. Some of these can be easily transformed so that a straight line can be drawn, others cannot.

The exponential graph

This graph follows a line where $y = a^x$

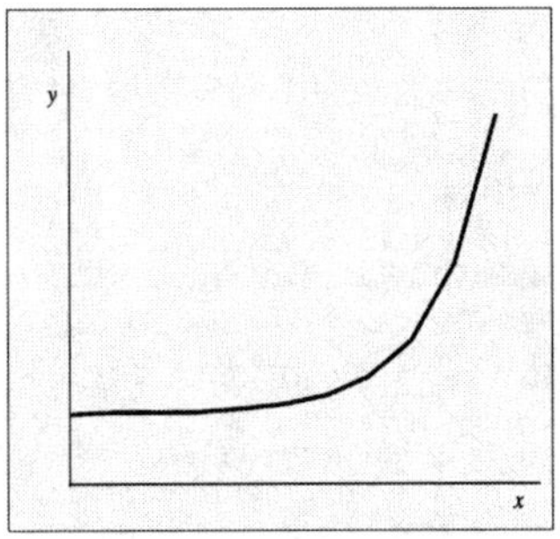

Figure 12.13a
An exponential curve

This is the equation for exponential growth where a is a constant. It is applicable to populations such as bacteria or yeast where individuals divide into two at regular intervals.

For example, where $a = 2$

$$y = 2^x$$

x	y
0	1
1	2
2	4
3	8
4	16
5	32

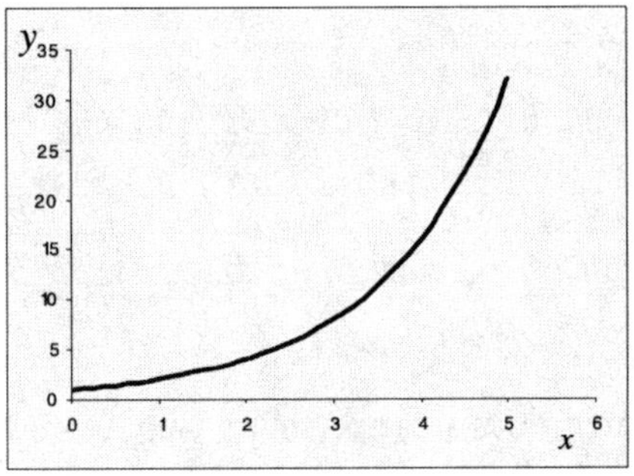

Figure 12.13b
Curve where $y=2^x$

In populations that grow exponentially, this constant is approximately 2.718...... It is a number that occurs frequently in relation to natural phenomena. If a population follows this exponential growth:

$$y = 2.718^x$$

x	y
0	1
1	2.718
2	7.388
3	20.079
4	54.576
5	148.336

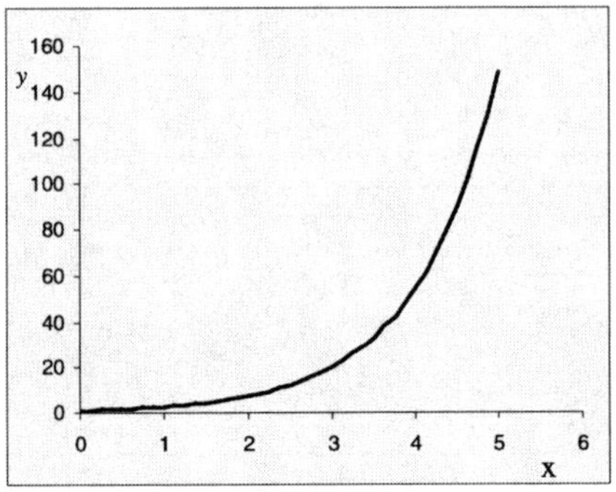

Figure 12.13c
Curve where $y=2.718^x$

A biological situation that uses this relationship is where population growth is calculated.

The equation used for determining the growth of an exponentially growing population is:

$$N = N_0 e^{kt}$$

N= the number in the population
N_0 = the number in the population when $t = 0$
k = growth or decay constant
$e = 2.718$

This is plotted on a graph:

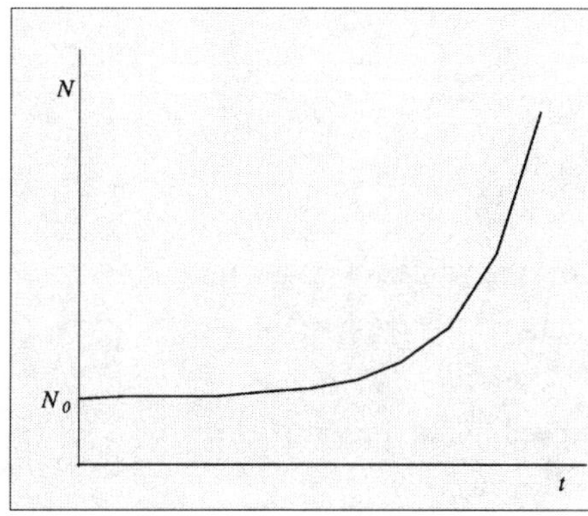

Figure 12.13d
An exponentially-growing population

This equation can be rearranged using logarithms (see chapter on Logarithms), in this case log to the base e, referred to as ln.

$$\ln N - \ln N_0 = kt$$
$$\ln N = kt + \ln N_0$$

This is now in the form $y = mx + c$

So y is plotted against x to give a straight line graph, in this case ln N is plotted against t.

From this straight line plot the gradient, m, can be calculated. This value is k in the above equation.

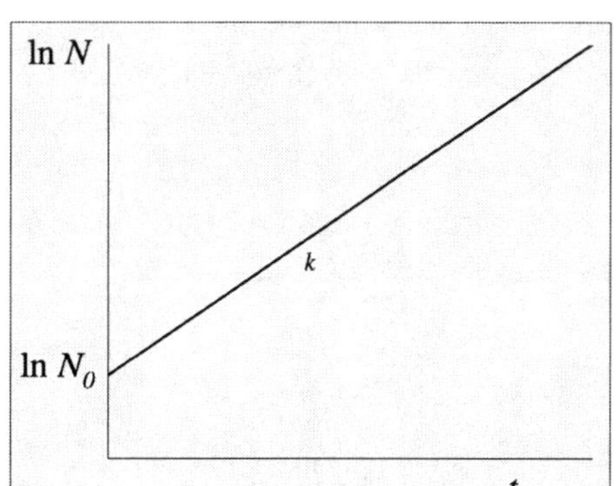

Figure 12.13e
Straight line plot of an exponentially-growing population

The quadratic graph

Quadratic graphs follow the equation $y = ax^2 + bx + c$

If your data follow a line that suggests a quadratic relationship, there will be key points of the graph that will be of interest. These are the points where the line crosses the x-axis, the intercept on the y-axis and the maximum or minimum value of y at the point of inflection.

If your data follows a line that has an equation:

$$y = x^2 + 6x - 8$$

It will look like this graph:

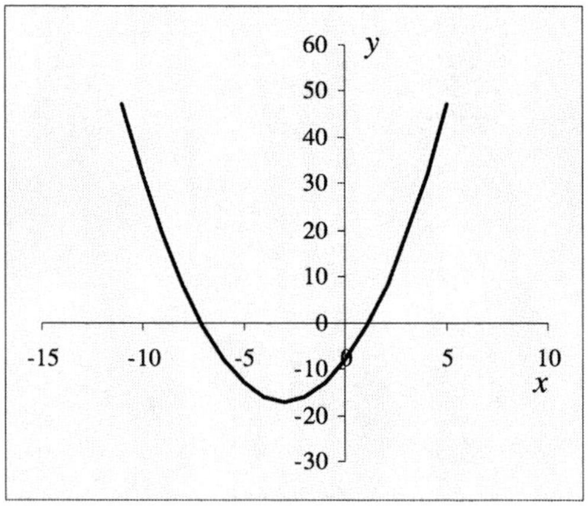

Figure 12.13f
Quadratic graph

It is often useful to know the values of x where the line crosses the x-axis, i.e. where $y = 0$.

To determine this you need to put the values for the equation into the following:

$$x = \frac{-b \pm \sqrt{b^2 - 4ac}}{2a}$$

where, $a = 1$, $b = +6$, $c = -8$

First substitute the letters for the numbers in the equation:

$$x = \frac{-6 \pm \sqrt{36 - (4 \times 1 \times -8)}}{2 \times 1}$$

Then simplify:

$$x = \frac{-6 \pm \sqrt{36 + 32}}{2}$$

$$x = \frac{-6 \pm 8.24}{2}$$

This provides two values for x:

$$x = -7.12$$

$$x = +1.12$$

If your graph follows the equation :

$$y = x^2 - 6x + 8$$

The graph will look like this:

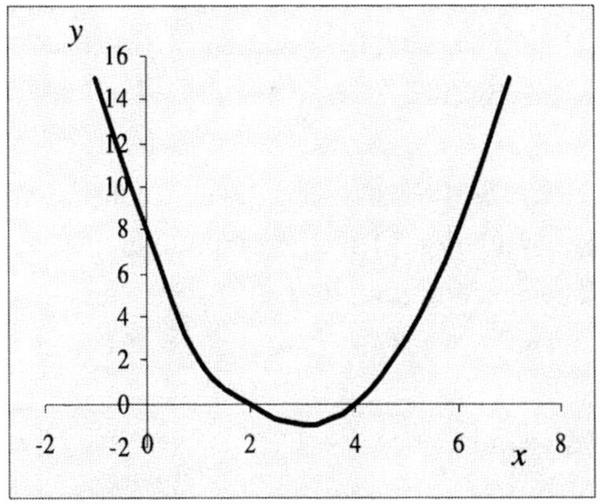

Figure 12.13g
A quadratic graph that factorises easily

In this case the equation can be factorised by eye (see chapter on Factorisation) to give:

$$y = (x - 4)(x - 2)$$

The curve crosses the x-axis where $y = 0$, so at this point:

$$0 = (x - 4)(x - 2)$$

For y to equal zero in this equation, one of the terms in brackets must equal zero, because any value multiplied by zero, will be equal to zero. So you can take each expression in brackets and solve it for x, given that it must equal zero.

$$x - 4 = 0 \qquad\qquad x - 2 = 0$$

$$x = +4 \qquad\qquad x = +2$$

So the line crosses the x-axis at +2 and +4.

The minimum point of the curve will be midway between these two points, for this graph it will be where $x = +3$. If this value for x is put into the equation for the graph, you can determine the minimum value of y.

$$y = x^2 - 6x + 8 \qquad y = 3^2 - (6 \times 3) + 8 \qquad y = 9 - (18) + 8 \qquad y = -1$$

Always look back at the graph and estimate the values for x and y at the minimum point, and check that the answer you get is close to these values. See the chapter on Calculus for a method of obtaining these values when the numbers in the equation of the graph are more complex.

There are only a few biological examples of the quadratic graph with a minimum value. One is the relationship between the isoelectric point of casein and pH (see chapter on Introduction to Calculus).

There are more examples of data that might follow a quadratic graph with a maximum value. For example, you could obtain a data set that followed this type of curve if you were to study a microbial population that required a certain amount of a substance to grow, but the substance became toxic at higher concentrations.

The points at which the line crosses the x-axis could be the concentrations of the substance that would give zero growth of the microbial population. The highest point on the graph could be the maximum growth of the microbial population. The concentration of the substance that gives this maximum growth would be equidistant between the two values of x, where the line crosses the x-axis.

An example of this type of quadratic graph could be: $y = -(x^2 + 6x - 8)$

This equation would produce the following graph.

Note that the only difference between this equation and the equation that refers to Figure12.13f is that the a term, the number in front of x^2, is a negative number, in this case it is -1.

If a quadratic equation has a negative number in front of the x^2, it will always be a curve with a maximum value.

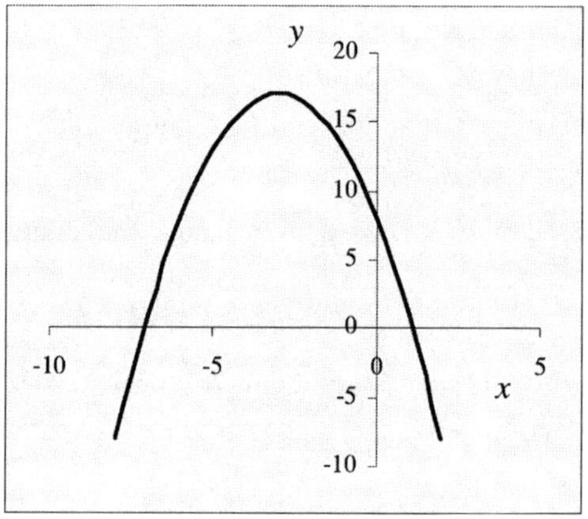

Figure 12.13h Quadratic curve with a maximum value

When the equation of this curve is put into the following $x = \dfrac{-b \pm \sqrt{b^2 - 4ac}}{2a}$ where $a = -1$, $b = -6$, $c = +8$

$$x = \frac{+6 \pm \sqrt{-6^2 - (4 \times -1 \times 8)}}{2 \times -1} \qquad x = \frac{+6 \pm \sqrt{36 + (32)}}{-2}$$

$$x = \frac{+6 \pm 8.24}{-2} \qquad x = -7.12 \qquad x = +1.12$$

You should note that this is not a good model for biological optima, it is usually just a description of a shape which data might follow.

To understand the quadratic equation, it is useful to look at what happens to the shape of graphs as different parts of the equation change. In the previous example you can see what happens when there is a negative sign in front of the x^2 part of the equation. It does not matter whether the negative sign is only in front of the x^2, or in front of the whole equation, as in the previous example. The only thing that causes the graph to have a maximum value rather than a minimum value, is having a negative value for a.

Remember the quadratic equation is the form $y = ax^2 + bx + c$

Looking at what happens as a,b and c change:

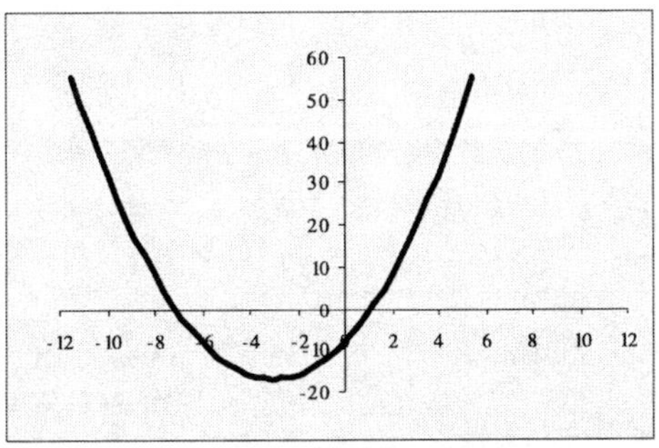

$$y = x^2 + 6x - 8$$

Take this graph as the one to compare the other three to.

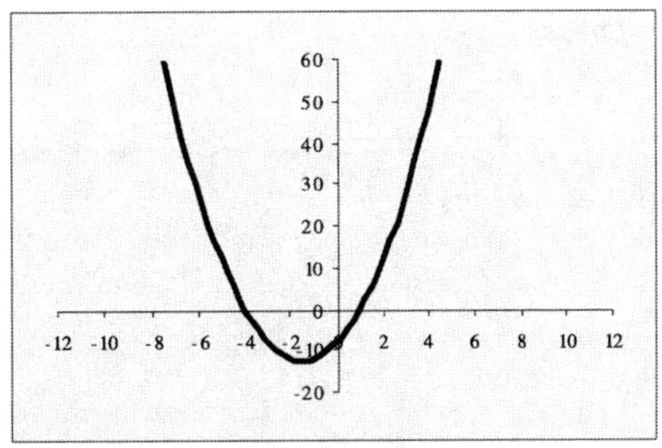

$$y = 2x^2 + 6x - 8$$

In this graph the value of a has increased. This has the effect of narrowing the overall shape of the curve.

Figure 12.13i Four quadratic graphs with different equations

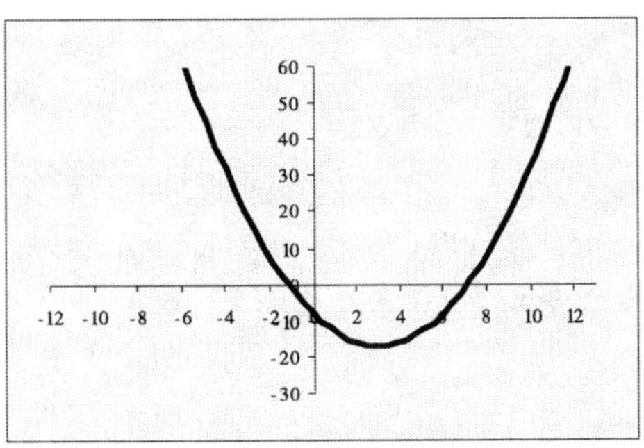

$$y = x^2 - 6x - 8$$

In this graph the value for b is negative. This has shifted the whole curve to the right so that the minimum part of the curve has a positive value on the x axis.

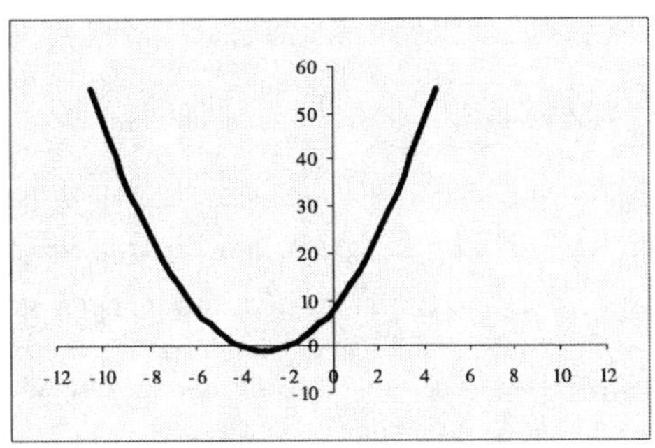

$$y = x^2 + 6x + 8$$

Here the value for c has increased and is a positive number. c is a constant, not linked to x, and its value is the point where the line crosses the y axis. In this case, increasing the value of c causes the graph to have a higher minimum value.

For further treatment of quadratic graphs see the chapter on Calculus.

The cubic graph

Equations that follow the formula

$$y = ax^3 + bx^2 + cx + d$$

are S-shaped, i.e. there are two points of inflection where the line changes direction. Each value in the equation, *a*, *b*, *c* and *d*, all have different effects on the shape of the curve as they change.

The value *d* is a constant that is not linked to *x*. As with the quadratic graph this value is equal to the point that the curve crosses the *y* axis.

As with the quadratic graph, each constant affects the shape and position of the graph on the axis. Below are two examples with different values for *b*.

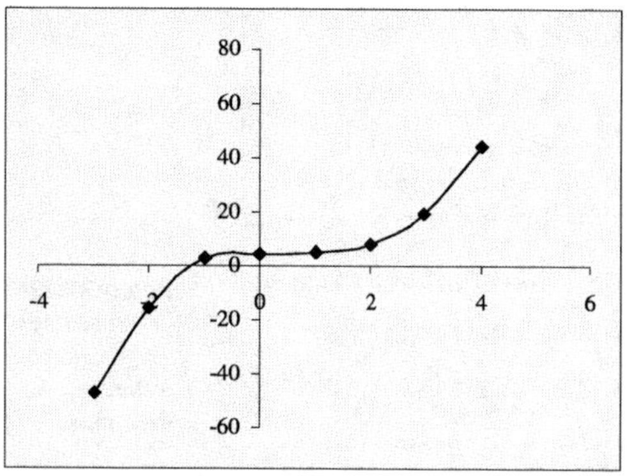

$$x^3 - 2x^2 + 2x + 4 = y$$

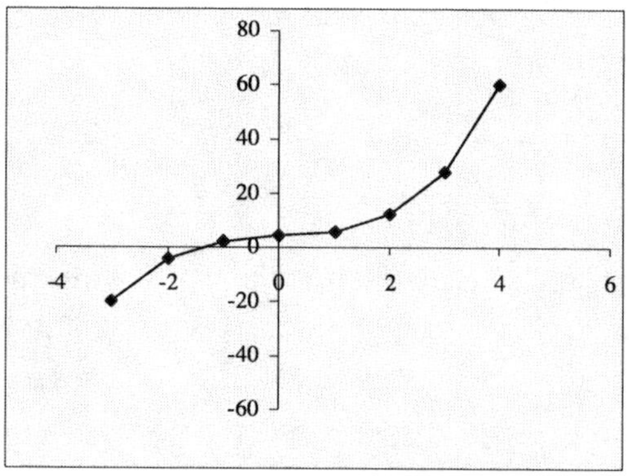

$$x^3 - x^2 + 2x + 4 = y$$

Figure 12.13j
S-shaped curves

A word of caution here about this type of plot. If you were to produce some data like this below, you can ask a spreadsheet software package to put a polynomial regression line through your data.

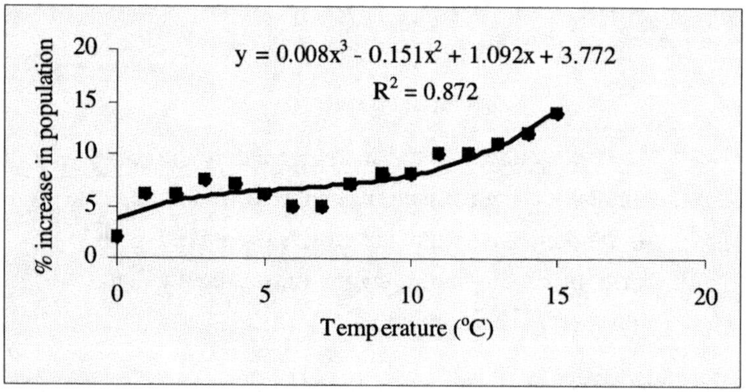

Figure12.13k
A third order polynomial regression line

The equation for the line in Figure 12.13k is given on the graph, with an R^2 value of 0.872. The R^2 value tells you how close the points are to the regression line; an indication of the confidence you might have in whether or not the regression line shows a correct relationship between the data. The closer R^2 is to 1, the closer the points are to the chosen regression line.

You can get a higher R^2 value by increasing the order of the polynomial regression. See what happens if you ask the spreadsheet software package to fit a fourth order polynomial line to the data.

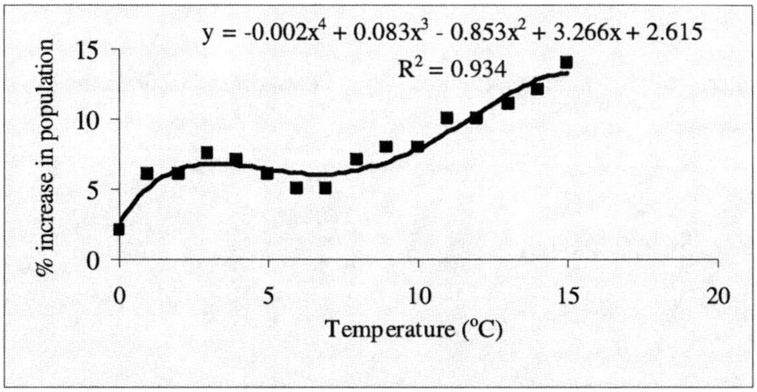

Figure12.13l
A fourth order polynomial regression line

The regression line has more points of inflection in it, and fits more closely to the data points, indicated by the R^2 value of 0.934. Now compare this to fitting a linear regression line through the data.

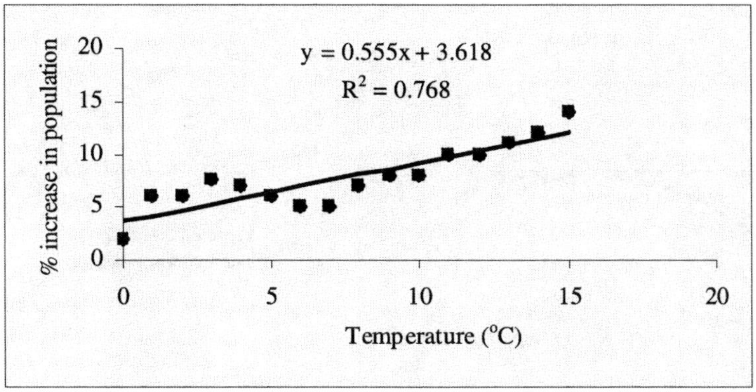

Figure 12.13m
A linear regression line

The R^2 value might suggest that a polynomial line is the one that best describes the relationship of the data. However, if you use a polynomial plot in this way, you must be able to justify it biologically, not statistically. Most biological situations you will come across will be linear, so would merit a straight line regression plot, as in Figure 12.13m. As with all data, you must think about the biology over and above the statistics.

The hyperbolic, or asymptotic, curve

You might meet this type of graph in several different areas of biology. The line of this graph is linear, up to a certain point, then starts to curve, and reaches its maximum point at infinity. It is at this point that the graph approaches a horizontal, straight line but does not reach it, and this is called an asymptote.
The equation that gives this shaped graph can be rearranged to give a straight line plot of the data. This is very useful as it allows to you obtain key data points from a straight line plot, that would not be accurately obtained from the non-linear plot.

You will come across this type of graph when studying enzyme kinetics. The equation used to describe the action of enzymes on a substrate, or enzyme kinetics is:

$$v = \frac{V_{max}[S]}{K_m + [S]}$$ This is in the form: $$y = \frac{ax}{b + x}$$

This is the Michaelis-Menten equation.

Where v = the speed of the enzyme reaction
V_{max} is the line which gives the maximum value for v which the line of the graph never reaches
K_m is the amount of substrate that gives half the maximum rate of reaction
$[S]$ is the substrate concentration

If this is plotted on a graph of v against $[S]$ you get the following:

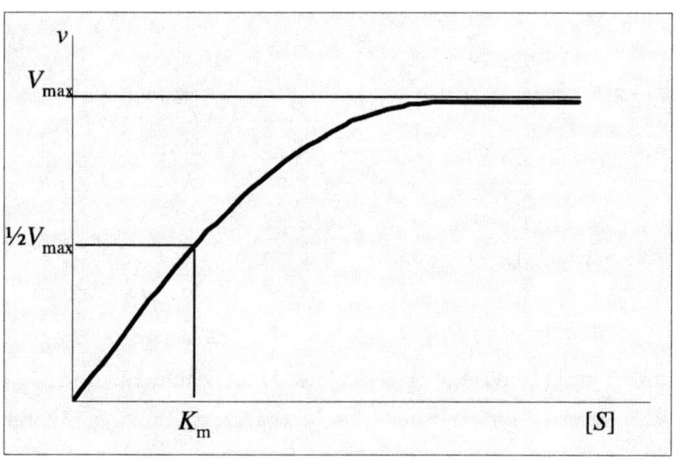

Figure 12.13n
The hyperbolic curve

At low substrate concentrations, when $[S]$ is less than K_m, the relationship is linear.
At high substrate concentrations, when $[S]$ is greater than K_m, the line begins to curve.
The line of the graph levels off but is never constant: it reaches V_{max} at infinity.

The equation that gives you this asymptotic graph can be rearranged so that the same data can be plotted as a straight line. There are several ways of doing this. The most common is the Lineweaver-Burk plot where $\frac{1}{V}$ is plotted against $\frac{1}{[S]}$ (To see how the equation is rearranged, see the chapter on Transposing Formulae.)

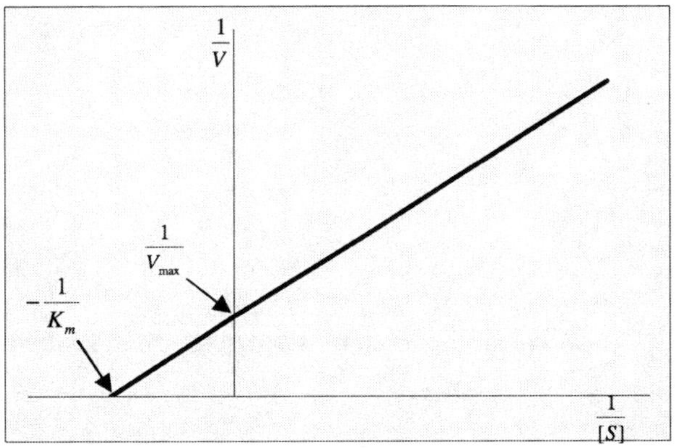

Figure 12.13o
The Lineweaver-Burk plot

You can see this if you compare the Lineweaver-Burk plot to the equation for a straight-line graph:

$$\frac{1}{v} = \frac{K_m}{V_{max}} \frac{1}{[S]} + \frac{1}{V_{max}}$$

$$y = m \ x \ + c$$

There are other biological applications of this asymptotic graph. One is in population biology with the Monod equation which relates to microbial growth.

$$\mu = \mu_{max} \frac{s}{K_s + s}$$

Here μ is the specific growth rate of the population.
μ_{max} is the maximum specific growth rate that the population reaches at infinity.
s is the concentration of a substance that becomes limiting to growth.
K_s is the constant that is the concentration of the substance at half μ_{max}.

It can be rearranged in the same way as the Michaelis-Menten equation, to produce a straight-line plot.

Test your understanding

Sketch a graph where

5. $y = x^2 - 2$

6. $y = x^2 + 2x$

7. $y = 3^x$

8. $y = \frac{1}{x}$

In the following equations, which are transpositions of the Michaelis-Menten formula, what would you plot on the x and y axes to get straight line plots.

9. $v = -K_m \frac{v}{[S]} + V_{max}$

10. $\frac{[S]}{v} = \frac{[S]}{V_{max}} + \frac{K_m}{V_{max}}$

11. From the equation $\ln N = kt + \ln N_0$ what is the gradient in a plot of ln$N$ against t?

In an experiment looking at the growth of *Chara* with increasing copper concentration, the data were plotted and were found to follow a quadratic shape with the equation $y = -0.04x^2 + 4.07x - 12.70$ where y was the growth of *Chara* in μm d^{-1} and x was the concentration of copper in μg l^{-1}.

12. Does this data have a maximum or a minimum point?

13. At what concentrations of copper is there no growth of *Chara*?

14. What is the growth of *Chara* at a copper concentration of 80 μg l^{-1}?

Chapter 13
Logarithm

03

The logarithm function

Objectives: This chapter:

- explains the term 'base of a logarithm'
- shows how to calculate the logarithm of a number to any base
- states the laws of logarithms and uses them to simplify expressions
- shows how to solve exponential and logarithmic equations
- defines the logarithm function
- illustrates graphically the logarithm function

Introducing logarithms

Given an equation such as $125 = 5^3$, we call 5 the base and 3 the power or index. We can use **logarithms** to write the equation in another form. The logarithm form is

$$\log_5 125 = 3$$

This is read as 'logarithm to the base 5 of 125 is 3'. In general if

$$y = a^x$$

then

$$\log_a y = x$$

Key point $y = a^x$ and $\log_a y = x$ are equivalent.

'The logarithm to the base a of y is x' is equivalent to saying that 'y is a to the power x'. The word logarithm is usually shortened to just 'log'.

WORKED EXAMPLE

13.1 Write down the logarithm form of the following:

(a) $16 = 4^2$ (b) $8 = 2^3$ (c) $25 = 5^2$

Solution (a) $16 = 4^2$ may be written as

$$2 = \log_4 16$$

that is, 2 is the logarithm to the base 4 of 16.

(b) $8 = 2^3$ may be expressed as

$$3 = \log_2 8$$

which is read as '3 is the logarithm to the base 2 of 8'.

(c) $25 = 5^2$ may be expressed as

$$2 = \log_5 25$$

that is, 2 is the logarithm to the base 5 of 25.

Given an equation such as $16 = 4^2$ then 'taking logs' to base 4 will result in the logarithmic form $\log_4 16 = 2$.

WORKED EXAMPLE

13.2 Write the exponential form of the following:

(a) $\log_2 16 = 4$ (b) $\log_3 27 = 3$ (c) $\log_5 125 = 3$ (d) $\log_{10} 100 = 2$

Solution (a) Here the base is 2 and so we may write $16 = 2^4$.

(b) The base is 3 and so $27 = 3^3$.

(c) The base is 5 and so $125 = 5^3$.

(d) The base is 10 and so $100 = 10^2$.

Recall that e is the constant 2.71828 ..., which occurs frequently in natural phenomena.

Although the base of a logarithm can be any positive number other than 1, the commonly used bases are 10 and e. Logarithms to base 10 are often denoted by 'log' or '$\log_{10}$'; logarithms to base e are denoted by 'ln' or '$\log_e$' and referred to as **natural logarithms**. Most scientific calculators possess 'log' and 'ln' buttons, which are used to evaluate logarithms to base 10 and base e.

WORKED EXAMPLES

13.3 Use a scientific calculator to evaluate the following:

(a) $\log 71$ (b) $\ln 3.7$ (c) $\log 0.4615$ (d) $\ln 0.5$ (e) $\ln 1000$

Solution Using a scientific calculator we obtain

(a) $\log 71 = 1.8513$ (b) $\ln 3.7 = 1.3083$ (c) $\log 0.4615 = -0.3358$
(d) $\ln 0.5 = -0.6931$ (e) $\ln 1000 = 6.9078$

You should ensure that you know how to use your calculator to verify these results.

13.4 Given $10^{0.6990} = 5$ evaluate the following:

(a) $10^{1.6990}$ (b) $\log 5$ (c) $\log 500$

Solution (a) $10^{1.6990} = 10^1.10^{0.6990} = 10(5) = 50$

(b) We are given $10^{0.6990} = 5$ and by taking logs to the base 10 we obtain $0.6990 = \log 5$.

(c) From $10^{0.6990} = 5$ we can see that

$$100(5) = 100(10^{0.6990})$$

and so

$$500 = 10^2.10^{0.6990} = 10^{2.6990}$$

Taking logs to the base 10 gives

$$\log 500 = 2.6990$$

Self-assessment question 13.1

1. The base of a logarithm is always a positive integer. True or false?

Exercise 13.1

1. Use a scientific calculator to evaluate
(a) $\log 150$ (b) $\ln 150$
(c) $\log 0.316$ (d) $\ln 0.1$

2. Write down the logarithmic form of the following:
(a) $3^8 = 6561$ (b) $6^5 = 7776$
(c) $2^{10} = 1024$ (d) $10^5 = 100000$
(e) $4^7 = 16384$
(f) $\left(\frac{1}{2}\right)^5 = 0.03125$
(g) $12^3 = 1728$ (h) $9^4 = 6561$

3. Write the exponential form of the following:
(a) $\log_6 1296 = 4$ (b) $\log_{15} 225 = 2$
(c) $\log_8 512 = 3$ (d) $\log_7 2401 = 4$
(e) $\log_3 243 = 5$ (f) $\log_6 216 = 3$
(g) $\log_{20} 8000 = 3$ (h) $\log_{16} 4096 = 3$
(i) $\log_2 4096 = 12$

4. Given $10^{0.4771} = 3$, evaluate the following:
(a) $\log 3$ (b) $\log 300$ (c) $\log 0.03$

5. Given $\log 7 = 0.8451$ evaluate the following:
(a) $10^{0.8451}$ (b) $\log 700$ (c) $\log 0.07$

Calculating logarithms to any base

On page 220 we showed how equations of the form $^x y = a$ could be written in logarithmic form as $\log_a y = x$. The number a, called the **base**, is always positive, that is, $a > 0$. We also introduced the common bases 10 and e. Scientific calculators are programmed with logarithms to base 10 and to base e. Suppose that we wish to calculate logarithms to bases other than 10 or e. For example, in communication theory, logarithms to base 2 are commonly used. To calculate logarithms to base a we use one of the following two formulae:

Key point

$$\log_a X = \frac{\log_{10} X}{\log_{10} a} \qquad \log_a X = \frac{\ln X}{\ln a}$$

WORKED EXAMPLE

13.5 Evaluate (a) $\log_6 19$, (b) $\log_7 29$.

Solution (a) We use the formula

$$\log_a X = \frac{\log_{10} X}{\log_{10} a}$$

Comparing $\log_6 19$ with $\log_a X$ we see $X = 19$ and $a = 6$. So

$$\log_6 19 = \frac{\log_{10} 19}{\log_{10} 6} = \frac{1.2788}{0.7782} = 1.6433$$

We could have equally well used the formula

$$\log_a X = \frac{\ln X}{\ln a}$$

With this formula we obtain

$$\log_6 19 = \frac{\ln 19}{\ln 6} = \frac{2.9444}{1.7918} = 1.6433$$

(b) Comparing $\log_a X$ with $\log_7 29$ we see $X = 29$ and $a = 7$. Hence

$$\log_7 29 = \frac{\log_{10} 29}{\log_{10} 7} = \frac{1.4624}{0.8451} = 1.7304$$

Alternatively we use

$$\log_7 29 = \frac{\ln 29}{\ln 7} = \frac{3.3673}{1.9459} = 1.7304$$

By considering the formula

$$\log_a X = \frac{\log_{10} X}{\log_{10} a}$$

with $X = a$ we obtain

$$\log_a a = \frac{\log_{10} a}{\log_{10} a} = 1$$

Key point

$\log_a a = 1$

This same result could be derived by writing the logarithm form of $a = a^1$.

Exercise 13.2

1. Evaluate the following:
(a) $\log_4 6$ (b) $\log_3 10$ (c) $\log_{20} 270$ (d) $\log_5 0.65$ (e) $\log_2 100$
(f) $\log_2 0.03$ (g) $\log_{100} 10$ (h) $\log_7 7$

2. Show that

$$2.3026 \log_{10} X = \ln X$$

3. Evaluate the following
(a) $\log_3 7 + \log_4 7 + \log_5 7$ (b) $\log_8 4 + \log_8 0.25$ (c) $\log_{0.7} 2$ (d) $\log_2 0.7$

Exponentials and logarithms in biology

What is a logarithm?

A logarithm is a very clever system which takes a number and represents that number as a power to which a particular base is raised to.

For example:

100 can be represented as 10 raised to the power of 2, or $100 = 10^2$

Here the base is 10. Keeping the base as 10, you can see how the number alters as 10 is raised to different powers.

10 raised to the power of 2.4 is 251 or $10^{2.4} = 251$

10 raised to the power of 3.6 is 3981 or $10^{3.6} = 3981$

10 raised to the power of 5.2 is 158489 or $10^{5.2} = 158489$

So you can take the number 100 and rewrite it as 10^2, the number 251 can be rewritten as $10^{2.4}$, the number 3981 can be rewitten as $10^{3.6}$ and the number 158489 can be rewritten as $10^{5.2}$.
Notice that the numbers range from 100 to 158,489 but the powers, to which 10 is raised, range from 2 to 5.2.

Imagine these numbers are areas of forests, in hectares, representing different sizes of protected habitat. The numbers of jaguars found in each area are 2, 3, 5 and 7 respectively. If you had to represent this on graph of jaguar numbers against forest area, you would get the following:

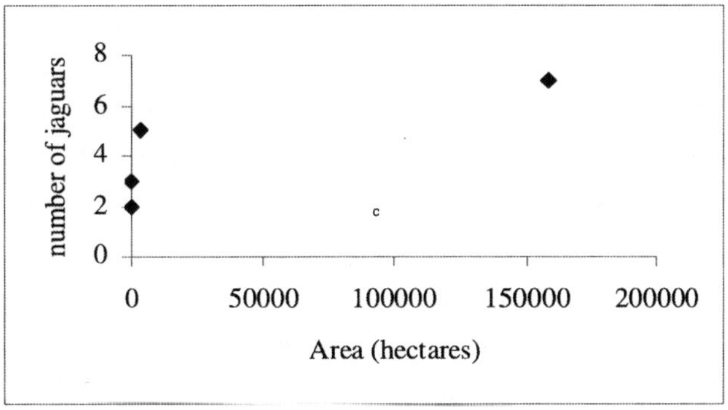

Fig 13.01
Jaguar numbers against forest area

The problem with this graph (Fig 13.01) is that most of the data are squashed in one tiny bit of the graph. Instead of the number of hectares of each area of forest, you could represent this as the number to which 10 is raised to give each area. So 10^2, $10^{2.4}$, $10^{3.6}$ and $10^{5.2}$ could be plotted as 2, 2.4, 3.6 and 5.2:

Fig 13.02
Jaguar numbers plotted against the area represented as the power to which 10 is raised.

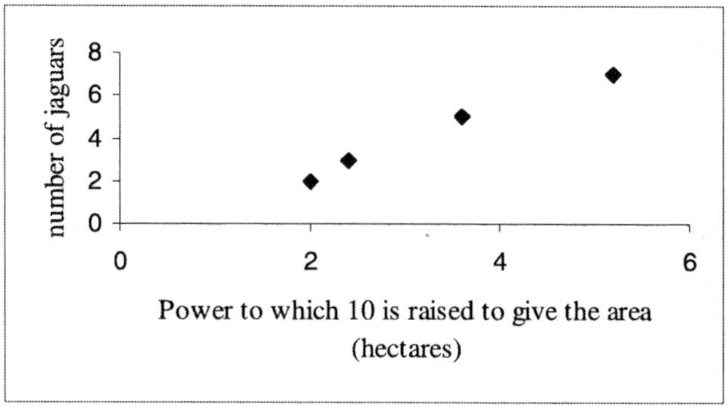

These powers are referred to as logarithms to the base 10. So by plotting the powers to which 10 is raised, you are actually plotting the logarithms of the numbers. So the x-axis of the graph can be referred to as the log of the area in hectares:

Fig 13.03
Jaguar numbers against the logarithm of the forest area

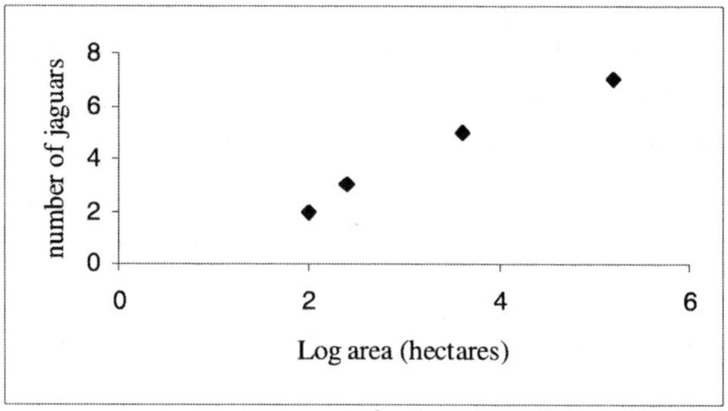

Using logarithms does two key things to the data. Firstly it transforms data that form a curve into a linear relationship. The reason it changes from a curve to a straight line is that logarithms are used to change a power equation into a linear equation. This can be seen clearly when looking at how you convert between logarithms and exponentials (see below). Also see Figure 13.1.
Secondly, it spreads the data more evenly along the x-axis. This can be useful in many situations. For example, you may need to represent a range of organisms on an axis, where the organisms range from bacteria to extremely large animals and plants.

Table 13.1 Standard vs logarithm scale of size of various organisms

	Size	Size in metres	Logarithm of size
Bacterium	0.2 μm	2×10^{-7}	-6.699
Amoeba	0.6 mm	6×10^{-4}	-3.222
Ant	4 mm	4×10^{-3}	-2.398
Mouse	12 cm	1.2×10^{-1}	-0.921
Horse	2 m	2	0.301
Whale	12 m	12	1.079
Giant sequoia tree	120 m	120	2.079

From this table you can see clearly that a scale that ranges from 2×10^{-7} to 120 m, can be transformed by logarithms so that the scale now ranges from -6.699 to 2.079. These data can then be presented on a scale on a graph or diagram that allows you to make comparisons between very small and very large organisms.

Use the table above to check that you can convert between numbers and logarithms using your calculator.

Logarithms are also very useful in representing a series of powers, for example:

$$10^1 = 10 \qquad \log 10 = 1$$

$$10^2 = 10 \times 10 = 100 \qquad \log 100 = 2$$

$$10^3 = 10 \times 10 \times 10 = 1000 \qquad \log 1000 = 3$$

$$10^4 = 10 \times 10 \times 10 \times 10 = 10000 \qquad \log 10000 = 4$$

And so on……..

So you can see clearly from this series, how the logarithm is equal to the power to which 10 is raised.

Converting between logarithms and exponentials

Take the expression:

$$10^{2.4} = 251$$

Here the power to which 10 has to be raised, to give 251, is 2.4 or the logarithm to the base 10 of 251 is 2.4.

The logarithm is the power to which the base is raised. However, the base does not have to be 10; it can be any number. Calculators are programmed so that the button labelled 'log' calculates logarithms to the base 10, but this is because it is a base that is often used.

For example: $2^3 = 8$

Here the base is 2, and if 2 is raised to the power 3, it equals 8 (2 x 2 x 2).

This can be rewritten as the logarithm to the base 2 of 8 is 3, or $\log_2 8 = 3$

Another example: $3^4 = 81$

The base is 3 and it is raised to the power 4, which is 3x3x3x3 = 81

Using logarithms it can be written as: $\log_3 81 = 4$ Here the logarithm to the base 3 of 81 is 4.

For any such equation you can use the following:

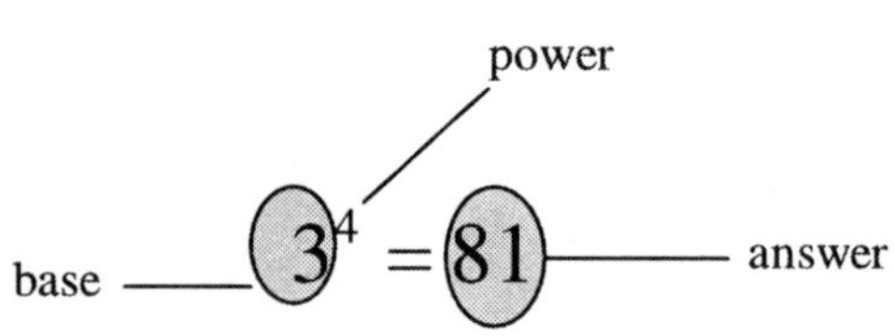

Log to the base of the answer equals the power.

Using this you can transform any exponential equation into a logarithmic equation.

When transforming a logarithmic equation into an exponential equation it is useful to remember that the answer of the equation is the power to which the base is raised.

Test your understanding....

Write these exponential expressions in their logarithmic form:

1. $2^4 = 16$
2. $3^5 = 243$
3. $5^2 = 25$
4. $8^3 = 512$
5. $4^3 = 64$

Write these logarithmic equations in their exponential form:

6. $\log_2 8 = 3$
7. $\log_9 81 = 2$
8. $\log_3 243 = 5$
9. $\log_4 64 = 3$
10. $\log_7 2401 = 4$

Fill in the missing number:

11. $10^5 = ??$
12. $\log_2 ?? = 2$
13. $\log_{??} 64 = 3$
14. $??^3 = 27$
15. $4^{??} = 16$

Laws of logarithms

Logarithms obey several laws, which we now examine. They are introduced via examples.

WORKED EXAMPLE

13.6 Evaluate (a) $\log 7$, (b) $\log 12$, (c) $\log 84$ and $\log 7 + \log 12$. Comment on your findings.

Solution (a) $\log 7 = 0.8451$ (b) $\log 12 = 1.0792$
(c) $\log 84 = 1.9243$, and $\log 7 + \log 12 = 0.8451 + 1.0792 = 1.9243$

We note that $\log 7 + \log 12 = \log 84$.

Example 13.6 illustrates the **first law** of logarithms, which states:

Key point

$$\log A + \log B = \log AB$$

This law holds true for any base. However, in any one calculation all bases must be the same.

WORKED EXAMPLES

13.7 Simplify to a single log term

(a) $\log 9 + \log x$
(b) $\log t + \log 4t$
(c) $\log 3x^2 + \log 2x$

Solution
(a) $\log 9 + \log x = \log 9x$
(b) $\log t + \log 4t = \log(t.4t) = \log 4t^2$
(c) $\log 3x^2 + \log 2x = \log(3x^2.2x) = \log 6x^3$

13.8 Simplify

(a) $\log 7 + \log 3 + \log 2$
(b) $\log 3x + \log x + \log 4x$

Solution (a) We know $\log 7 + \log 3 = \log(7 \times 3) = \log 21$, and so

$$\log 7 + \log 3 + \log 2 = \log 21 + \log 2$$
$$= \log(21 \times 2) = \log 42$$

(b) We have

$$\log 3x + \log x = \log(3x.x) = \log 3x^2$$

and so

$$\log 3x + \log x + \log 4x = \log 3x^2 + \log 4x$$
$$= \log(3x^2.4x) = \log 12x^3$$

We now consider an example that introduces the second law of logarithms.

WORKED EXAMPLE

13.9 (a) Evaluate $\log 12$, $\log 4$ and $\log 3$.
(b) Compare the values of $\log 12 - \log 4$ and $\log 3$.

Solution (a) $\log 12 = 1.0792, \log 4 = 0.6021, \log 3 = 0.4771$.

(b) From part (a),

$$\log 12 - \log 4 = 1.0792 - 0.6021 = 0.4771$$

and also

$$\log 3 = 0.4771$$

We note that $\log 12 - \log 4 = \log 3$.

This example illustrates the **second law** of logarithms, which states:

Key point

$$\log A - \log B = \log\left(\frac{A}{B}\right)$$

WORKED EXAMPLES

13.10 Use the second law of logarithms to simplify the following to a single log term:

(a) $\log 20 - \log 10$ (b) $\log 500 - \log 75$ (c) $\log 4x^3 - \log 2x$
(d) $\log 5y^3 - \log y$

Solution (a) Using the second law of logarithms we have

$$\log 20 - \log 10 = \log\left(\frac{20}{10}\right) = \log 2$$

(b) $$\log 500 - \log 75 = \log\left(\frac{500}{75}\right) = \log\left(\frac{20}{3}\right)$$

(c) $$\log 4x^3 - \log 2x = \log\left(\frac{4x^3}{2x}\right) = \log 2x^2$$

(d) $$\log 5y^3 - \log y = \log\left(\frac{5y^3}{y}\right) = \log 5y^2$$

13.11 Simplify

(a) $\log 20 + \log 3 - \log 6$
(b) $\log 18 - \log 24 + \log 2$

Solution (a) Using the first law of logarithms we see that

$$\log 20 + \log 3 = \log 60$$

and so

$$\log 20 + \log 3 - \log 6 = \log 60 - \log 6$$

Using the second law of logarithms we see that

$$\log 60 - \log 6 = \log\left(\frac{60}{6}\right) = \log 10$$

Hence

$$\log 20 + \log 3 - \log 6 = \log 10$$

(b)
$$\begin{aligned}\log 18 - \log 24 + \log 2 &= \log\left(\frac{18}{24}\right) + \log 2\\ &= \log\left(\frac{3}{4}\right) + \log 2\\ &= \log\left(\frac{3}{4} \times 2\right)\\ &= \log 1.5\end{aligned}$$

13.12 Simplify

(a) $\log 2 + \log 3x - \log 2x$
(b) $\log 5y^2 + \log 4y - \log 10y^2$

Solution (a)
$$\begin{aligned}\log 2 + \log 3x - \log 2x &= \log(2 \times 3x) - \log 2x\\ &= \log 6x - \log 2x\\ &= \log\left(\frac{6x}{2x}\right) = \log 3\end{aligned}$$

(b)
$$\begin{aligned}\log 5y^2 + \log 4y - \log 10y^2 &= \log(5y^2.4y) - \log 10y^2\\ &= \log 20y^3 - \log 10y^2\\ &= \log\left(\frac{20y^3}{10y^2}\right) = \log 2y\end{aligned}$$

We consider a special case of the second law. Consider $\log A - \log A$. This is clearly 0. However, using the second law we may write

$$\log A - \log A = \log\left(\frac{A}{A}\right) = \log 1$$

Thus

Key point

$$\log 1 = 0$$

In any base, the logarithm of 1 equals 0.

Finally we introduce the third law of logarithms.

WORKED EXAMPLE

13.13 (a) Evaluate $\log 16$ and $\log 2$.
(b) Compare $\log 16$ and $4 \log 2$.

Solution (a) $\log 16 = 1.204, \log 2 = 0.301$.
(b) $\log 16 = 1.204$, $4 \log 2 = 1.204$. Hence we see that $4 \log 2 = \log 16$.

Noting that $16 = 2^4$, example 13.13 suggests the **third law** of logarithms:

Key point

$$n \log A = \log A^n$$

This law applies if n is integer, fractional, positive or negative.

WORKED EXAMPLES

13.14 Write the following as a single logarithmic expression:

(a) $3 \log 2$ (b) $2 \log 3$ (c) $4 \log 3$

Solution (a) $3 \log 2 = \log 2^3 = \log 8$
(b) $2 \log 3 = \log 3^2 = \log 9$
(c) $4 \log 3 = \log 3^4 = \log 81$

13.15 Write as a single log term

(a) $\frac{1}{2}\log 16$ (b) $-\log 4$ (c) $-2 \log 2$ (d) $-\frac{1}{2}\log 0.5$

Solution (a) $\frac{1}{2}\log 16 = \log 16^{1/2} = \log \sqrt{16} = \log 4$

(b) $-\log 4 = -1.\log 4 = \log 4^{-1} = \log\left(\frac{1}{4}\right) = \log 0.25$

(c) $-2\log 2 = \log 2^{-2} = \log\left(\frac{1}{2^2}\right) = \log\left(\frac{1}{4}\right) = \log 0.25$

(d) $-\frac{1}{2}\log 0.5 = -\frac{1}{2}\log\left(\frac{1}{2}\right) = \log\left(\frac{1}{2}\right)^{-1/2} = \log 2^{1/2} = \log\sqrt{2}$

13.16 Simplify

(a) $3\log x - \log x^2$
(b) $3\log t^3 - 4\log t^2$
(c) $\log Y - 3\log 2Y + 2\log 4Y$

Solution (a) $3\log x - \log x^2 = \log x^3 - \log x^2$

$$= \log\left(\frac{x^3}{x^2}\right)$$

$$= \log x$$

(b) $3\log t^3 - 4\log t^2 = \log(t^3)^3 - \log(t^2)^4$

$$= \log t^9 - \log t^8$$

$$= \log\left(\frac{t^9}{t^8}\right)$$

$$= \log t$$

(c) $\log Y - 3\log 2Y + 2\log 4Y = \log Y - \log(2Y)^3 + \log(4Y)^2$

$$= \log Y - \log 8Y^3 + \log 16Y^2$$

$$= \log\left(\frac{Y.16Y^2}{8Y^3}\right)$$

$$= \log 2$$

13.17 Simplify

(a) $2\log 3x - \frac{1}{2}\log 16x^2$

(b) $\frac{3}{2}\log 4x^2 - \log\left(\frac{1}{x}\right)$

(c) $2\log\left(\frac{2}{x^2}\right) - 3\log\left(\frac{2}{x}\right)$

Solution (a) $2\log 3x - \frac{1}{2}\log 16x^2 = \log(3x)^2 - \log(16x^2)^{1/2}$

$$= \log 9x^2 - \log 4x$$

$$= \log\left(\frac{9x^2}{4x}\right)$$

$$= \log\left(\frac{9x}{4}\right)$$

(b) $\frac{3}{2}\log 4x^2 - \log\left(\frac{1}{x}\right) = \log(4x^2)^{3/2} - \log(x^{-1})$

$$= \log 8x^3 + \log x$$

$$= \log 8x^4$$

(c) $2\log\left(\frac{2}{x^2}\right) - 3\log\left(\frac{2}{x}\right) = \log\left(\frac{2}{x^2}\right)^2 - \log\left(\frac{2}{x}\right)^3$

$$= \log\left(\frac{4}{x^4}\right) - \log\left(\frac{8}{x^3}\right)$$

$$= \log\left(\frac{4/x^4}{8/x^3}\right)$$

$$= \log\left(\frac{1}{2x}\right)$$

Self-assessment question 13.2

1. State the three laws of logarithms.

Exercise 13.3

1. Write the following as a single log term using the laws of logarithms:
   (a) $\log 5 + \log 9$ (b) $\log 9 - \log 5$
   (c) $\log 5 - \log 9$ (d) $2\log 5 + \log 1$
   (e) $2\log 4 - 3\log 2$ (f) $\log 64 - 2\log 2$
   (g) $3\log 4 + 2\log 1 + \log 27 - 3\log 12$

2. Simplify as much as possible:
   (a) $\log 3 + \log x$
   (b) $\log 4 + \log 2x$
   (c) $\log 3X - \log 2X$
   (d) $\log T^3 - \log T$
   (e) $\log 5X + \log 2X$

3. Simplify
(a) $3\log X - \log X^2$ (b) $\log y - 2\log\sqrt{y}$
(c) $5\log x^2 + 3\log\frac{1}{x}$
(d) $4\log X - 3\log X^2 + \log X^3$
(e) $3\log y^{1.4} + 2\log y^{0.4} - \log y^{1.2}$
(h) $4\log\sqrt{x} + 2\log\left(\frac{1}{x}\right)$
(i) $2\log x + 3\log t$ (j) $\log A - \frac{1}{2}\log 4A$
(k) $\frac{\log 9x + \log 3x^2}{3}$

4. Simplify the following as much as possible by using the laws of logarithms:
(a) $\log 4x - \log x$ (b) $\log t^3 + \log t^4$
(c) $\log 2t - \log\left(\frac{t}{4}\right)$
(d) $\log 2 + \log\left(\frac{3}{x}\right) - \log\left(\frac{x}{2}\right)$
(e) $\log\left(\frac{t^2}{3}\right) + \log\left(\frac{6}{t}\right) - \log\left(\frac{1}{t}\right)$
(f) $2\log y - \log y^2$
(g) $3\log\left(\frac{1}{t}\right) + \log t^2$
(l) $\log xy + 2\log\left(\frac{x}{y}\right) + 3\log\left(\frac{y}{x}\right)$
(m) $\log\left(\frac{A}{B}\right) - \log\left(\frac{B}{A}\right)$
(n) $\log\left(\frac{2t}{3}\right) + \frac{1}{2}\log 9t - \log\left(\frac{1}{t}\right)$

5. Express as a single log term:
$\log_{10} X + \ln X$

6. Simplify
(a) $\log(9x - 3) - \log(3x - 1)$
(b) $\log(x^2 - 1) - \log(x + 1)$
(c) $\log(x^2 + 3x) - \log(x + 3)$

Solving equations with logarithms

This section illustrates the use of logarithms in solving certain types of equations. For reference we note the equivalence of

$$y = a^x \text{ and } \log_a y = x$$

and the laws of logarithms:

$$\log A + \log B = \log AB$$

$$\log A - \log B = \log\left(\frac{A}{B}\right)$$

$$n\log A = \log A^n$$

WORKED EXAMPLES

13.18 Solve the following equations:

(a) $10^x = 59$ (b) $10^x = 0.37$ (c) $e^x = 100$ (d) $e^x = 0.5$

Solution

(a) $10^x = 59$
Taking logs to base 10 gives

$$x = \log 59 = 1.7709$$

(b) $10^x = 0.37$
Taking logs to base 10 gives

$$x = \log 0.37 = -0.4318$$

(c) $e^x = 100$
Taking logs to base e gives

$$x = \ln 100 = 4.6052$$

(d) $e^x = 0.5$
Taking logs to base e we have

$$x = \ln 0.5 = -0.6931$$

13.19 Solve the following equations:

(a) $\log x = 1.76$ (b) $\ln x = -0.5$ (c) $\log(3x) = 0.76$

(d) $\ln\left(\frac{x}{2}\right) = 2.6$ (e) $\log(2x - 4) = 1.1$ (f) $\ln(7 - 3x) = 1.75$

Solution

We note that if $\log_a X = n$, then $X = a^n$.

(a) $\log x = 1.76$ and so $x = 10^{1.76}$. Using the 'x^y' button of a scientific calculator we find

$$x = 10^{1.76} = 57.5440$$

(b) $\ln x = -0.5$

$$x = e^{-0.5} = 0.6065$$

(c) $\log 3x = 0.76$

$$3x = 10^{0.76}$$

$$x = \frac{10^{0.76}}{3} = 1.9181$$

(d) $\ln\left(\frac{x}{2}\right) = 2.6$

$$\frac{x}{2} = e^{2.6}$$

$$x = 2e^{2.6} = 26.9275$$

(e) $\log(2x - 4) = 1.1$

$$2x - 4 = 10^{1.1}$$

$$2x = 10^{1.1} + 4$$

$$x = \frac{10^{1.1} + 4}{2} = 8.2946$$

(f) $\ln(7 - 3x) = 1.75$

$$7 - 3x = e^{1.75}$$

$$3x = 7 - e^{1.75}$$

$$x = \frac{7 - e^{1.75}}{3} = 0.4151$$

13.20 Solve the following:

(a) $e^{3+x}.e^{x} = 1000$ (b) $\ln\left(\frac{x}{3} + 1\right) + \ln\left(\frac{1}{x}\right) = -1$

(c) $3e^{x-1} = 75$ (d) $\log(x + 2) + \log(x - 2) = 1.3$

Solution (a) Using the laws of logarithms we have

$$e^{3+x}.e^{x} = e^{3+2x}$$

Hence

$$e^{3+2x} = 1000$$

$$3 + 2x = \ln 1000$$

$$2x = \ln(1000) - 3$$

$$x = \frac{\ln(1000) - 3}{2} = 1.9539$$

(b) Using the laws of logarithms we may write

$$\ln\left(\frac{x}{3}+1\right)+\ln\left(\frac{1}{x}\right)=\ln\left(\left(\frac{x}{3}+1\right)\frac{1}{x}\right)=\ln\left(\frac{1}{3}+\frac{1}{x}\right)$$

So

$$\ln\left(\frac{1}{3}+\frac{1}{x}\right)=-1$$

$$\frac{1}{3}+\frac{1}{x}=e^{-1}$$

$$\frac{1}{x}=e^{-1}-\frac{1}{3}$$

$$\frac{1}{x}=0.0345$$

$$x=28.947$$

(c) $3e^{x-1}=75$

$$e^{x-1}=25$$

$$x-1=\ln 25$$

$$x=\ln 25+1=4.2189$$

(d) Using the first law we may write

$$\log(x+2)+\log(x-2)=\log(x+2)(x-2)=\log(x^2-4)$$

Hence

$$\log(x^2-4)=1.3$$

$$x^2-4=10^{1.3}$$

$$x^2=10^{1.3}+4$$

$$x=\sqrt{10^{1.3}+4}=4.8941$$

Exercise 13.4

1. Solve the following equations, giving your answer to 4 d.p.:
(a) $\log x=1.6000$ (b) $10^x=75$
(c) $\ln x=1.2350$ (d) $e^x=36$

2. Solve each of the following equations, giving your answer to 4 d.p.:
(a) $\log(3t)=1.8$ (b) $10^{2t}=150$
(c) $\ln(4t)=2.8$ (d) $e^{3t}=90$

3. Solve the following equations, giving your answer to 4 d.p.:
   (a) $\log x = 0.3940$ (b) $\ln x = 0.3940$
   (c) $10^y = 5.5$ (d) $e^z = 500$
   (e) $\log(3v) = 1.6512$ (f) $\ln\left(\dfrac{t}{6}\right) = 1$
   (g) $10^{2r+1} = 25$ (h) $e^{(2t-1)/3} = 7.6700$
   (i) $\log(4b^2) = 2.6987$
   (j) $\log\left(\dfrac{6}{2+t}\right) = 1.5$
   (k) $\ln(2r^3 + 1) = 3.0572$
   (l) $\ln(\log t) = -0.3$ (m) $10^{t^2-1} = 180$
   (n) $10^{3r^4} = 170\,000$
   (o) $\log(10^t) = 1.6$ (p) $\ln(e^x) = 20\,000$
4. Solve the following equations:
   (a) $e^{3x}.e^{2x} = 59$
   (b) $10^{3t}.10^{4-t} = 27$
   (c) $\log(5-t) + \log(5+t) = 1.2$
   (d) $\log x + \ln x = 4$

Properties and graph of the logarithm function

Values of $\log x$ and $\ln x$ are given in Table 13.1, and graphs of the functions $y = \log x$ and $y = \ln x$ are illustrated in Figure 13.1. The following properties are noted from the graphs:

(a) As x increases, the values of $\log x$ and $\ln x$ increase.
(b) $\log 1 = \ln 1 = 0$.

Table 13.1

x	0.01	0.1	0.5	1	2	5	10	100
$\log x$	−2	−1	−0.30	0	0.30	0.70	1	2
$\ln x$	−4.61	−2.30	−0.69	0	0.69	1.61	2.30	4.61

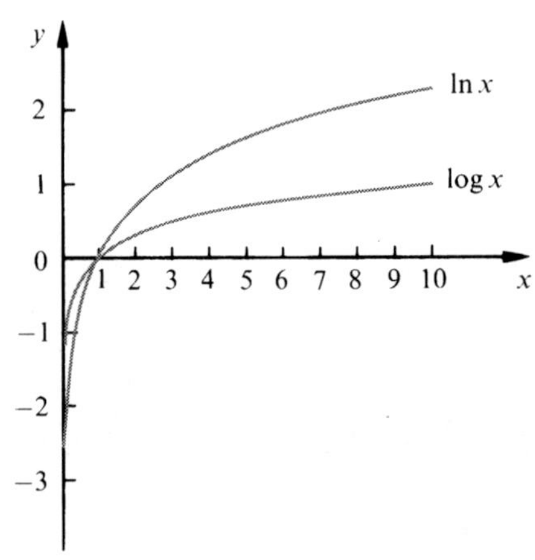

Figure 13.1
Graphs of $y = \log x$ and $y = \ln x$

(c) As x approaches 0 the values of $\log x$ and $\ln x$ increase negatively.
(d) When $x < 1$, the values of $\log x$ and $\ln x$ are negative.
(e) $\log x$ and $\ln x$ are not defined when $x \leqslant 0$.

Self-assessment questions 13.3

1. State two properties that are common to both $y = \log x$ and $y = \ln x$.
2. It is possible to find a value of x such that the value of $\log x$ exceeds 100 000 000. True or false?

Exercise 13.5

1. (a) Plot a graph of $y = \log x^2$ for $0 < x \leqslant 10$.
   (b) Plot a graph of $y = \log(1/x)$ for $0 < x \leqslant 10$.
2. (a) Plot a graph of $y = \log x$ for $0 < x \leqslant 3$.
   (b) Plot on the same axes $y = 1 - x/3$ for $0 \leqslant x \leqslant 3$.
   (c) Use your graphs to find an approximate solution to the equation

$$\log x = 1 - \frac{x}{3}$$

3. (a) Plot $y = \log x$ for $0 < x \leqslant 7$.
   (b) On the same axes, plot $y = 0.5 - 0.1x$.
   (c) Use your graphs to obtain an approximate solution to

$$0.5 - 0.1x = \log x$$

Test and assignment exercises 13

1. Use a scientific calculator to evaluate
   (a) $\log 107\,000$ (b) $\ln 0.0371$ (c) $\log 0.1$ (d) $\ln 150$
2. Evaluate
   (a) $\log_7 20$ (b) $\log_{16} 100$ (c) $\log_2 60$ (d) $\log_8 150$ (e) $\log_6 4$
3. Simplify the following expressions to a single log term:
   (a) $\log 7 + \log t$ (b) $\log t - \log x$ (c) $\log x + 2\log y$
   (d) $3\log t + 4\log r$ (e) $\frac{1}{2}\log 9v^4 - 3\log 1$ (f) $\log(a+b) + \log(a-b)$
   (g) $\frac{2}{3}\log x + \frac{1}{3}\log xy^3$

4. Solve the following logarithmic equations, giving your answer to 4 d.p.:
   (a) $\log 7x = 2.9$ (b) $\ln 2x = 1.5$ (c) $\ln x + \ln 2x = 3.6$
   (d) $\ln\left(\frac{x^2}{2}\right) - \ln x = 0.7$ (e) $\log(4t - 3) = 0.9$ (f) $\log 3t + 3\log t = 2$
   (g) $3\ln\left(\frac{2}{x}\right) - \frac{1}{2}\ln x = -1$

5. (a) Plot $y = \log x$ for $0 < x \leqslant 3$.
   (b) On the same axes plot $y = \sin x$, where x is measured in radians.
   (c) Hence find an approximate solution to $\log x = \sin x$.

6. Solve the following equations, giving your answer correct to 4 d.p.:
   (a) $10^{x+1} = 70$ (b) $e^{2x+3} = 500$ (c) $3(10^{2x}) = 750$ (d) $4(e^{-x+2}) = 1000$
   (e) $\frac{4}{6 + e^{3x}} = 0.1500$

7. By substituting $z = e^x$ solve the equation

   $$e^{2x} - 9e^x + 14 = 0$$

8. By substituting $z = 10^x$ solve the following equation, giving your answer correct to 4 d.p.:

   $$10^{2x} - 9(10^x) + 20 = 0$$

9. Solve the following equations:
   (a) $\log(x - 1) + \log(x + 1) = 2$ (b) $\log(10 + 10^x) = 2$ (c) $\frac{\log 2x}{\log x} = 2$

10. Solve
   (a) $8(10^{-x}) + 10^x = 6$ (b) $10^{3x} - 4(10^x) = 0$

Calculating pH

How acidic a solution is, is determined by the concentration of H^+ ions. The concentration of these ions ranges from very small (10^{-14} mol l^{-1}) to very large (1 mol l^{-1}). Because the scale is so large, logarithms can be used. The equation used is

$$pH = -\log_{10}\left[H^+\right]$$

The minus sign in front of the logarithm means that the pH number is positive, because the H^+ concentration is almost always less than 1. Here are some H^+ concentrations and their pH values.

Table 13.2
H^+ concentrations and their pH

$[H^+]$ (mol l^{-1})	equation	pH
0.1	-log [0.1]	1
10^{-7}	-log [10^{-7}]	7
4 x 10^{-10}	-log [4 x 10^{-10}]	9.398

For weak acids and bases, not all the water in the solution splits into H^+ and OH^- ions, so you cannot take the above pH equation to calculate pH from $[H^+]$ ions directly. The dissociation constant tells you how many H_2O molecules have split into H^+ and OH^- ions. The dissociation constant for an acid is K_a. You can determine K_a experimentally. The higher the value for K_a, the more H_2O has split and the more H^+ ions there are. As with pH, values can be very small and very large, so logarithms are taken.

For acids $$pK_a = -\log K_a$$

For bases $$pK_b = -\log K_b$$

Buffer systems are composed of an acid-base pair. Each acid-base combination has a specific pH. The pH of an acid-base pair can be calculated based on the formulae

$$pK_a = -\log K_a$$

$$pK_b = -\log K_b$$

A weak acid HA, splits into H+ and A- or

$$HA_{(aq)} + H_2O \Leftrightarrow A^-{}_{(aq)} + H_3O^+{}_{(aq)}$$

$$K_a = \frac{[A^-][H_3O^+]}{[HA]}$$ or it can be written as $$K_a = [H_3O^+] \times \frac{[A^-]}{[HA]}$$

Using the first law of logarithms this can be rewritten as

$$\log K_a = \log[H_3O^+] + \log[A^-] - \log[HA]$$

You then multiply through the equation by -1:

$$-\log K_a = -\log[H_3O^+] - \log[A^-] + \log[HA]$$

This then allows you to substitute the pK_a and pH equations

$$pK_a = -\log K_a$$

$$pH = -\log_{10}[H^+]$$

into this equation, which gives you:

$$pK_a = pH - \log[A^-] + \log[HA]$$

This equation $$pK_a = pH - \log\left[A^-\right] + \log\left[HA\right]$$

can then be rearranged further by taking pH across to the other side:

$$-pH + pK_a = -\log\left[A^-\right] + \log\left[HA\right]$$

And then taking pK_a across to the opposite side:

$$-pH = -pK_a - \log\left[A^-\right] + \log\left[HA\right]$$

The equation is then multiplied through by -1:

$$pH = pK_a + \log\left[A^-\right] - \log\left[HA\right]$$

This can be rewritten putting the positive logarithm on the top of a fraction, and a negative logarithm on the bottom of a fraction:

$$pH = pK_a + \log\frac{\left[A^-\right]}{\left[HA\right]}$$

This is called the **Henderson-Hasselbalch** equation.

It is worth noting that you don't often calculate pH when making up solutions, because you use a pH meter. However, it is really important to understand the mathematics behind the pH of solutions so that you fully understand what pH is.

Worked examples

[Acetic acid $pK_a = 4.7$]

1) An acetic acid/acetate buffer solution containing 0.2 mol l^{-1} acetic acid (acid or proton donor) and 0.05 mol l^{-1} sodium acetate (base or proton acceptor) would have a pH of 4.1

$$pH = pK_a + \log\frac{\left[A^-\right]}{\left[HA\right]}$$

$$pH = 4.7 + \log\left(\frac{0.05}{0.2}\right)$$

$$pH = 4.7 - 0.6$$

$$pH = 4.1$$

2) Calculate the pH of a mixture containing 100 ml of 0.1 mol l^{-1} acetic acid (CH_3CO_2H) and 300 ml of 0.2 mol l^{-1} sodium acetate ($NaCH_3CO_2$)?

In a question like this you must take into account dilution of each solution (100ml plus 300ml):

Concentration of acid (acetic acid CH_3CO_2H):

$$no.\,moles = 0.1 \times \frac{100}{1000} = 0.01\,moles$$

Total volume = 400ml, therefore

$$0.01\,moles = \frac{400}{1000} \times molarity$$

Molarity = 0.025 mol l^{-1}

By same reasoning (but note volumes) concentration of base (sodium acetate $NaCH_3CO_2$):

$$concentration\ of\ base = 0.2 \times \frac{300}{400} = 0.15\ mol\ l^{-1}$$

Putting these into Henderson-Hasselbalch equation :

$$pH = pK_a + \log\frac{[A^-]}{[HA]}$$

$$pH = 4.7 + \log\left(\frac{0.15}{0.025}\right)$$

$$pH = 4.7 + 0.8$$

$$pH = 5.5$$

Test your understanding

16. Calculate the pH of an acetic acid/acetate buffer solution containing 0.3 mol l^{-1} acetic acid (acid or proton donor) and 0.15 mol l^{-1} sodium acetate (base or proton acceptor).
[Acetic acid $pKa = 4.7$]

17. Calculate the pH of an acetic acid/acetate buffer solution containing 0.1 mol l^{-1} acetic acid (acid or proton donor) and 0.3 mol l^{-1} sodium acetate (base or proton acceptor).
[Acetic acid $pKa = 4.7$]

18. Calculate the pH of a mixture containing 200 ml of 0.1 mol l^{-1} acetic acid (CH_3CO_2H) and 150 ml of 0.2 mol l^{-1} sodium acetate ($NaCH_3CO_2$)?

19. Calculate the pH of a mixture containing 100 ml of 0.05 mol l^{-1} acetic acid (CH_3CO_2H) and 200 ml of 0.3 mol l^{-1} sodium acetate ($NaCH_3CO_2$)?

20. Calculate the pH of a 2 litre solution containing 10ml of 5 mol l^{-1} acetic acid and 10ml of 1 mol l^{-1} sodium acetate.
[Acetic acid $pKa = 4.7$]

21. The pK_a of formic acid is 3.77. What is the pH of a 0.02 mol l^{-1} solution of formic acid in water?

22. Tris is a weak base ($pK_a = 8.08$) commonly used as a buffer in biochemical systems. What is the ratio of the concentrations of tris and its conjugate acid at pH 8.0?

23. If an acid has a H^+ concentration of 0.00001 mol l^{-1}, what is the pH of the solution?

24. If an acid has a H^+ concentration of 0.00000001 mol l^{-1}, what is the pH of the solution?

25. If an acid has a H^+ concentration of 0.0005 mol l^{-1}, what is the pH of the solution?

26. If a base has an OH^- concentration of 0.001 mol l^{-1}, what is the pH? Assume that $[H^+][OH^-] = 10^{-14}$

27. If a base has an OH^- concentration of 0.01 mol l^{-1}, what is the pH? Assume that $[H^+][OH^-] = 10^{-14}$

28. If a base has an OH^- concentration of 0.02μmol l^{-1}, what is the pH? Assume that $[H^+][OH^-] = 10^{-14}$

Natural logarithms

In biology you will generally find only two bases are used:

1) log to the base 10

2) log to the base 2.718281828 ....

Why such a specific number? This number is referred to as '*e*'. It is a number that comes up a lot in mathematics. Historically it is a number that had to be discovered, in the same way that π had to be discovered. The number *e* is linked to numbers that tend to infinity. One of its very early uses was in calculating compound interest.

The number *e*, 2.718281828 .... continues for an infinite number of decimal places. The graph that is related to the number e is the rectangular hyperbola (Fig 13.2). With this graph the values of x and y continue to infinity. The lines of the graph almost, but never actually, touch the axes.

For this graph when x is high y is low, and when y is high x is low, so that $xy = 1$,

Or $y = \frac{1}{x}$ so where $y = 1, x = 1$

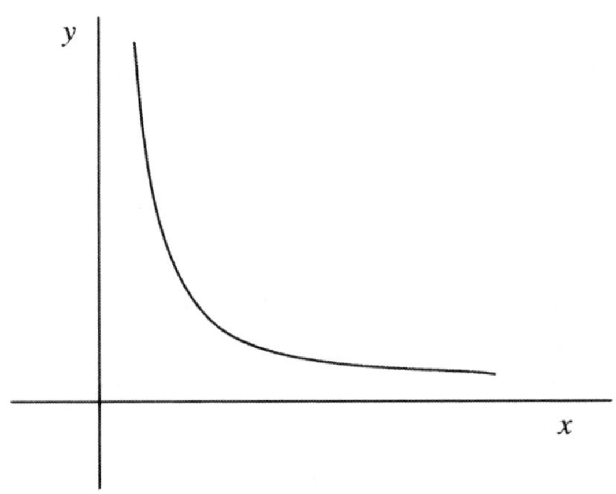

Fig 13.2
A rectangular hyperbola

This graph has certain properties that are crucial to logarithms and also to calculus, which is dealt with in the next chapter.

The area beneath the graph between where $x = 1$ and $x = 2$, is equal to ln 2.

The area under the graph between the points where $x = 1$ and $x = e$ is equal to 1.

It is the relationship between *x* and *y* on this graph (Fig 13.2) that makes *e* the base of natural logarithms. The key thing to understand is that *e* is a number that relates to numbers that continue to infinity, and that because of this it is a number that occurs a lot in biological processes. The root of *e,* in infinite series, means it is also a really important number in mathematics. Almost all calculations in mathematics and biology that include logarithms use natural logarithms, one exception being calculations of pH.

Calculators are not only programmed to calculate logarithms to the base 10 (log) they are also programmed to calculate logarithms to the base *e* (ln). Check you can use you calculator correctly for calculating logarithms to the base *e*:

$\ln 3 \times 10^4 = 10.309$
$\ln 4 \times 10^{-5} = -10.127$
$\ln 2.4 = 0.875$

You may come across the use of natural logarithms in biology when studying population dynamics, or calculating radioactive decay.

Examples of natural logarithms in biology

The equation used for determining the growth of an exponentially growing population is:

$$N = N_0 e^{kt}$$

N= the number in the population
N_0 = the number in the population when t=0
k = growth or decay constant
e = 2.718….

This equation is rearranged using logarithms, so that it becomes a linear equation. First it needs to be in the form of an answer, a base and a power, so:

$$\frac{N}{N_0} = e^{kt}$$

$\frac{N}{N_0}$ is the answer, e is the base ($\log_e$ is ln) and kt is the power.

So using logarithms this can now be rearranged:

$$\log_e \frac{N}{N_0} = kt \qquad \text{or} \qquad \ln \frac{N}{N_0} = kt$$

Using the second law of logarithms this equation can be expanded:

$$\ln N - \ln N_0 = kt$$

By rearranging this equation it can be rewritten in the form

$$y = mx + c$$

$$\ln N = kt + \ln N_0$$

For example, you can calculate the growth constant, k, of a population if you know the number that were present at time = 0, and the number after a specific time interval.

For example, if a population of 30 mice increased to a population of 125 in 160 days, what would the growth constant of this population be?

$N_0 = 30$ mice $\qquad t = 160$ days

$N = 125$ mice $\qquad k =$ growth constant

$$\ln \frac{N}{N_0} = kt \qquad \ln \frac{125}{30} = k \times 160 \qquad \ln 4.167 = k \times 160$$

$$1.427 = k \times 160 \qquad \frac{1.427}{160} = k \qquad 0.009\ \text{d}^{-1} = k$$

For another example, the numbers of children contracting measles were 756 in 2006. This number is predicted to double within one year. If the number of cases do double in one year, calculate the growth constant for this disease.

$N_0 = 1 \qquad t = 1$ year

$N = 2 \qquad k =$ growth constant

$$\ln \frac{N}{N_0} = kt \qquad \ln \frac{2}{1} = k \times 1 \qquad 0.69 = k \times 1$$

$$\frac{0.69}{1} = k \qquad 0.69\ \text{y}^{-1} = k$$

Another example where the equation: $N = N_0 e^{kt}$

is used in a biological situation, is when calculating decay constants of isotopes. In this case the constant k is referred to as the decay constant, and is a negative number.

For example, you may be asked to find the decay constant of an isotope with a half life of 6 hours.

For all such questions you can take it that the isotope would give a reading of 1 at time 0, or N_0, and, for the above question, six hours later the reading, *N*, would be 0.5, i.e. half what it was six hours earlier.

These values can be substituted into the above equation:

$0.5 = 1e^{k6}$ or $0.5 = e^{6k}$

Logarithms can be used to change this equation:

$\log_e 0.5 = 6k$ or $\ln 0.5 = 6k$

This can be rearranged to get *k* on its own:

$$\frac{\ln 0.5}{6} = k$$

$k = -0.116\ \mathrm{h}^{-1}$

k is negative because you are taking the logarithm of number less than one.

k is negative if it is a decay constant and positive if it is a growth constant.

Test your understanding.

Answer the following:

29. What is the doubling time of a population with an annual growth rate of 4.2%?
30. What is the growth rate of a population with a doubling time of 86 years?
31. What is the half life of an isotope with a decay constant of $-0.35\ \mathrm{day}^{-1}$?
32. What is the decay constant of an isotope with a half life of 30 s?

Chapter 14
An Introduction to Calculus

An Introduction to Calculus

Take the velocity of a moving object that moves at a constant speed in one direction. A graph of displacement against time will be a straight line with a gradient equal to the velocity of the object. To find the gradient you take the coordinates of two points on the line and use the equation:

$$gradient = \frac{difference\ between\ y\ coordinates}{difference\ between\ x\ coordinates}$$

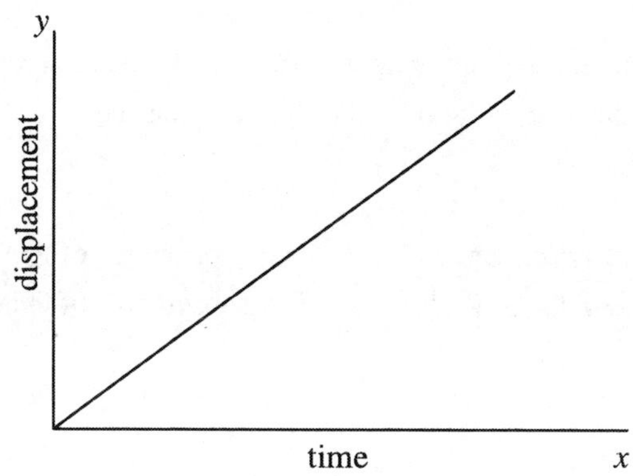

Fig 14.1 Displacement against time

Displacement is a function of time; the greater the length of time, the greater the displacement of the object.

So the value on the y-axis is a function of the value on the x-axis. The value on the y-axis can therefore be rewritten as $f(x)$, a function of x.

In more general terms, x is the independent variable, you choose the values of x. In contrast, y is the dependent variable, its value depends on the value of x, and what is done to it. So y can be referred to as the function of x, the function being what is done to x to give the value of y.

In the graph above, the line follows the equation $y = mx$

Multiplying by m, the gradient, is the operation which is done to x to give the value of y. So the equation of the line can be rewritten as:

$$f(x) = mx$$

If the equation of the line were $y = 2x$

the operation is 2; this is what is done to x to obtain the value of y, so

$$f(x) = 2x$$

Fig 14.2
Displacement against time showing functions of x

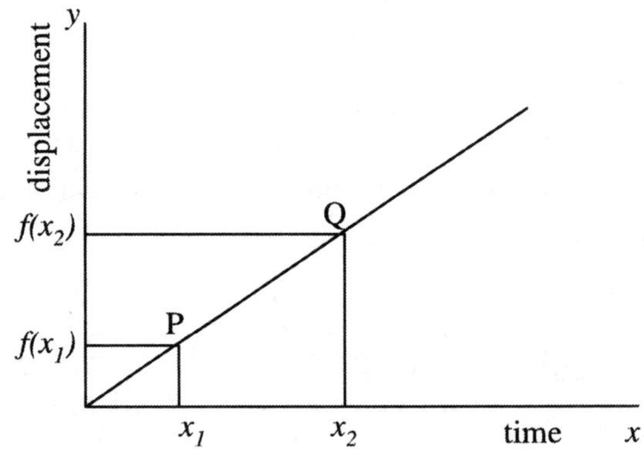

If you take the above graph, P is the point on the graph where $x = x_1$ and $y = f(x_1)$, and Q is the point where $x = x_2$ and $y = f(x_2)$. To calculate the gradient of the graph you use the equation:

$$gradient = \frac{difference\ between\ y\ coordinates}{difference\ between\ x\ coordinates}$$

The coordinates of P are $(x_1, f(x_1))$.
The coordinates of Q are $(x_2, f(x_2))$.

$$gradient = \frac{f(x_2) - f(x_1)}{x_2 - x_1}$$

Rather than using x, the values of x can be defined as different letters, so x_1 can become a, and x_2 can become b:

Fig 14.3
Displacement against time giving values to x

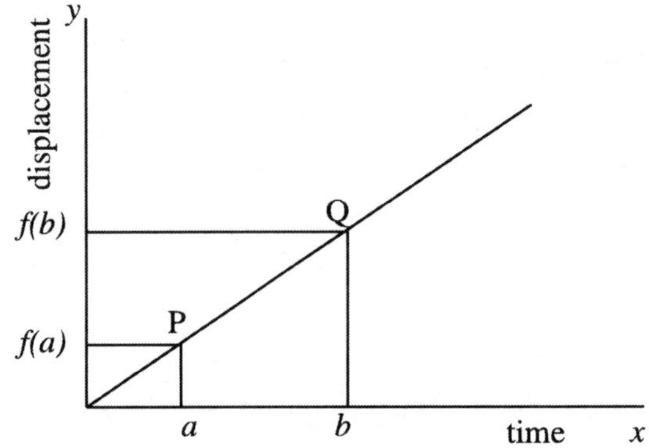

The coordinates of P are $(a, f(a))$.
The coordinates of Q are $(b, f(b))$.

So the gradient of the line can be calculated from these coordinates:

$$gradient = \frac{f(b) - f(a)}{b - a}$$

The velocity of an object can therefore be calculated by taking two points in time, and knowing the displacement of the object at each of these times.

Now, instead of taking two different times, you can take point *b* and call it point *a* plus a length of time, *h*. In other words:

$$b = a + h$$

Fig 14.4
Showing the difference between two values of *x*

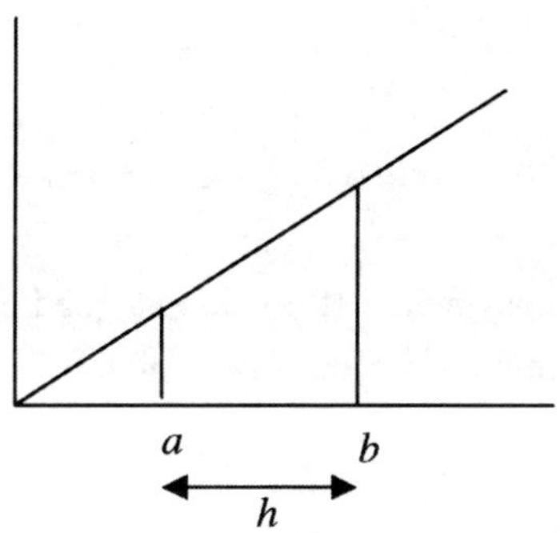

For example, if $a = 3$ min and $b = 5$ min,
then $h = 2$ min
because $b = a + h$
or $5 = 3 + 2$

So the equation from above:

$$gradient = \frac{f(b) - f(a)}{b - a}$$

can be rewritten substituting *a*+*h* for *b*:

$$gradient = \frac{f(a+h) - f(a)}{a + h - a}$$

and this can be simplified:

$$gradient = \frac{f(a+h) - f(a)}{h}$$

This equation can be used to describe the change in position of an object from position *f*(*a*) to position (*f*(*a*+*h*) over a period of time, *h*.

If we take a circle, and draw a line, t_1 , across the circle, it intersects the circle twice, at points Q and R below. This line is called a secant, or cord.

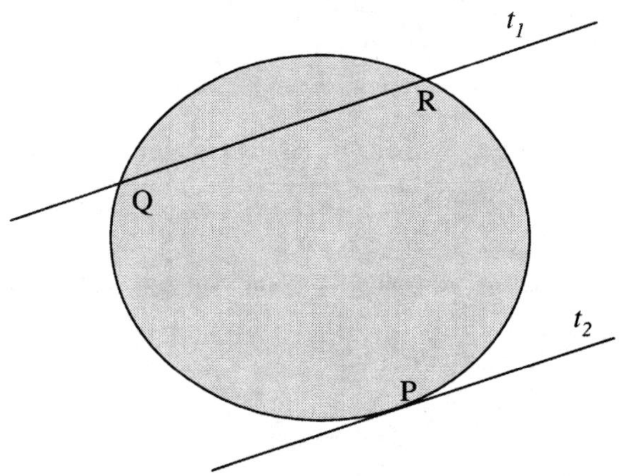

Fig 14.5
Showing a secant to a circle

Alternatively a line, t_2 , can be drawn that intersects the circle only once, at point P. This line is called a tangent.

This can also be applied to lines on graphs that are non-linear. If you take a quadratic graph you can draw a tangent to the point at which the graph is at its maximum (or minimum) point.

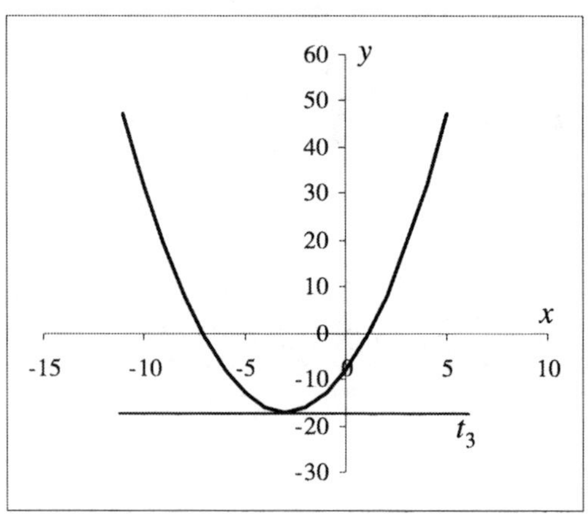

Fig 14.6
A tangent to a quadratic curve

Line t_3, the tangent to the curve, has a gradient that is equal to the gradient of the line at that point. The quadratic curve is at its minimum point where the gradient is equal to zero. At this point, the tangent is a line where, at different values of x, y always has the same value, so as x changes, y does not. The gradient at this point is therefo zero.

You can use the Power Rule to calculate the value of x at the point at which the gradient is equal to zero. The Power Rule is a formula that can be applied to any equation where y is a function of x.

For a given function $y = ax^n$ the Power Rule formula is anx^{n-1}

where a is the number in front of x and n is the power to which x is raised.

Take this graph:

Fig 14.7
Applying the Power Rule to the minimum value of a quadratic curve

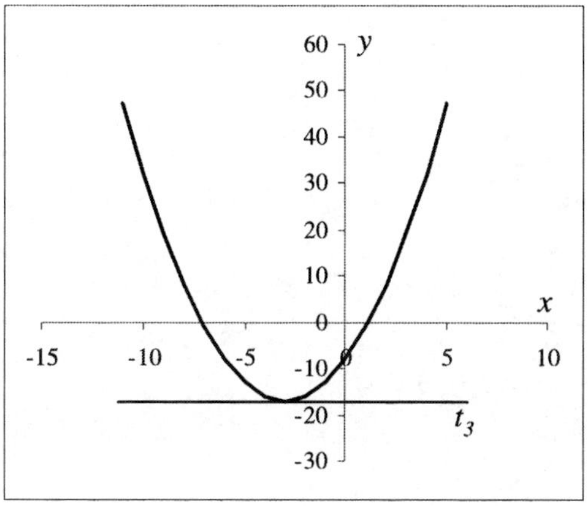

The equation of this graph is

$$y = x^2 + 6x - 8$$

This equation can be rewritten as:

$$y = 1x^2 + 6x^1 - 8x^0$$

In this form, each time x appears in the equation there is a value for a in front of x, and a value for n, the power to which x is raised.
Now, to each part of the equation, the Power Rule is applied.

Take the first part, $1x^2$, here $a = 1$ and $n = 2$, so substituting these numbers into anx^{n-1}

gives $1 \times 2x^{2-1}$

This can be simplified as

$$2x^1 \qquad \text{or} \qquad 2x$$

$$y = 1x^2 + 6x^1 - 8x^0$$

Take the second bit of the equation $6x^1$

and apply the Power Rule to this anx^{n-1}

$6 \times 1x^{1-1}$

and simplify it. $6x^0$ or 6.

Then take the third bit of the equation $8x^0$

and apply the Power Rule to it anx^{n-1}

$8 \times 0x^{0-1}$

and simplify it. $0x^{-1}$ or 0

Putting all this together: $1x^2 + 6x^1 - 8x^0$ becomes $2x + 6 - 0$

or $2x + 6$

This process relates to the gradient of the curve. When the gradient is equal to zero, when the quadratic graph is at its minimum value, the equation, to which the Power Rule has been applied, is equal to zero. This allows you to solve the equation for *x*.

$$2x + 6 = 0 \qquad 2x = -6 \qquad x = -\frac{6}{2} \qquad x = -3$$

This is the value for *x* when *y* is at its minimum value. To find the value for *y*, you can put this value for *x* back in the original equation:

$$y = x^2 + 6x - 8 \qquad y = -3^2 + (6 \times -3) - 8 \qquad y = 9 - 18 - 8 \qquad y = -17$$

How can this process be applied to a biological situation? Take a study to find the isoelectric point of casein, where the following data was obtained, and a graph plotted of pH against protein concentration (Fig 14.8). If you plot this data and estimate the minimum point by eye you would probably estimate a value for the protein concentration of around 0.1 mg ml^{-1} at a pH of around 4.4.

pH	Protein concentration (mg ml^{-1})
5.6	0.46
5.3	0.35
5.0	0.25
4.7	0.14
4.4	0.10
4.1	0.15
3.8	0.30
3.5	0.45

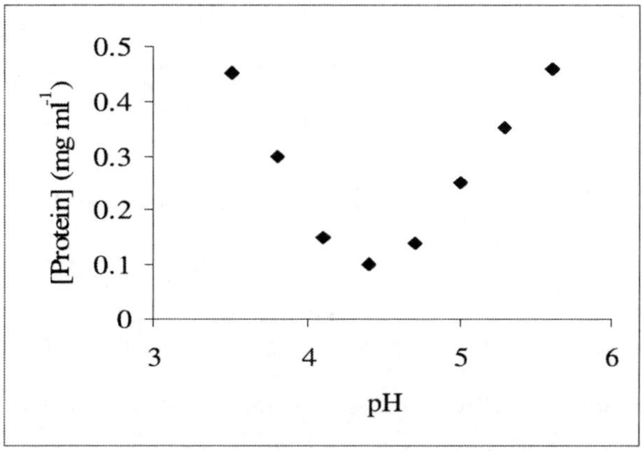

Fig 14.8 The isoelectric point of casein

Alternatively you could enter the data into a spreadsheet software package and ask the programme to add a quadratic line to the data (polynomial 2, or x^2). You also need the programme to display the equation on the graph.

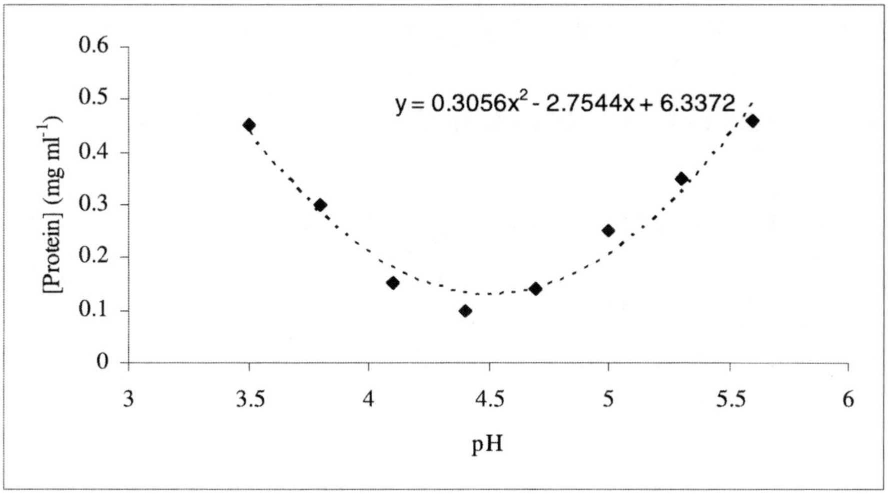

Fig 14.8 Using an equation of the curve to find the isoelectric point of casein

You then apply the Power Rule to this equation $y = 0.3056x^2 - 2.7544x + 6.3372$, as before.

$$y = 0.3056x^2 - 2.7544x + 6.3372$$

This can be rewritten as

$$y = 0.3056x^2 - 2.7544x^1 + 6.3372x^0$$

Applying anx^{n-1} to each part of the above equation:

$$0.3056 \times 2x^{2-1} - 2.7544 \times 1x^{1-1} + 6.3372 \times 0x^{0-1}$$

and simplifying: $$0.6112x^1 - 2.7544x^0 + 0x^{-1}$$

$$0.6112x \quad - 2.7544$$

At the point at which the curve is at its minimum point, the gradient is equal to zero, and this equation can be equal to zero, and solved for x.

$$0.6112x \quad - 2.7544 = 0 \qquad 0.6112x = +2.7544 \qquad x = \frac{2.7544}{0.6112} \qquad x = 4.5065$$

The pH value for which the protein concentration is at its lowest is 4.5. To find the minimum protein concentration, this value can be put into the original equation:

$$y = 0.3056x^2 - 2.7544x + 6.3372$$

$$y = 0.3056 \times 4.5^2 - 2.7544 \times 4.5 + 6.3372$$

$$y = 0.3056 \times 4.5^2 - 2.7544 \times 4.5 + 6.3372$$

$$y = 0.1308$$

The minimum protein concentration is 0.13 mg ml^{-1}. While you can estimate these values, this method uses mathematics to provide more accurate values. You must note, however, that the values are only as accurate as the data you originally collected, and the significant figures you quote must reflect the accuracy of the original data.

Test your understanding

Apply the Power Rule to the following equations:

1. $y = x^2 + 3x - 4$

2. $y = 2x^2 - 5x - 8$

3. $y = x^3 + 2x^2 - 8$

4. $y = 3x^{-2}$

5. $y = 4x^3 - 12$

6. $y = 4x^{-3}$

7. $y = x^2 - 4x + 6$

8. $y = x^3 + 3x$

9. $y = 4x^2 - 3x$

10. $y = x + 1$

11. Plot the following data in a spreadsheet software package. It gives the growth of rhizoids of *Chara* at different concentrations of copper.
Determine the concentration of copper that provides the maximum growth of this plant, and the maximum growth at this copper concentration.

Copper concentration (μg l^{-1})	Rhizoid growth (μm d^{-1})
0	0
10	10
20	50
30	80
40	80
50	100
60	100
70	80
80	60
90	60
100	0

Calculus in Biology

Differential calculus is the mathematical approach to finding the gradient (derivative) of continuous functions. For example: If a population, *N*, is growing in an exponential fashion with time according to the relationship,

$$N = N_o e^{kt}$$

where N_0 is the population at time 0, and *k* is the growth rate constant, then the differential of *N* with respect to time, $\frac{dN}{dt}$, is the gradient to the exponential curve and tells us the rate of change of the population at any given time, t.

For a function of a single variable, $f(x)$, the derivative at a point equals the slope of the tangent line to the graph of the function at that point.

The process of finding a derivative is called differentiation. The fundamental theorem of calculus states that differentiation is the reverse process to integration.

Differentiation has applications in many areas of study. The velocity of a moving body is the derivative of its displacement with respect to time and the derivative of velocity with respect to time is acceleration. The rate of a chemical reaction is a derivative.

Differentiation is frequently used to find the maxima and minima of a function. Equations involving derivatives are called differential equations and are fundamental in describing natural phenomena such as the rate of growth of a population.

Tangents

The word *tangent* is derived from the Latin word *tangens*, meaning "touching"

A tangent to a curve is a line that touches the curve. In other words, a tangent line should have the same slope as the curve at the point of contact. The slope tells us how rapidly the function is changing at that point i.e. its gradient.

What if we say that the tangent is a line that intersects the curve once and once only? Unfortunately, except for a circle this definition is generally inadequate (see below):

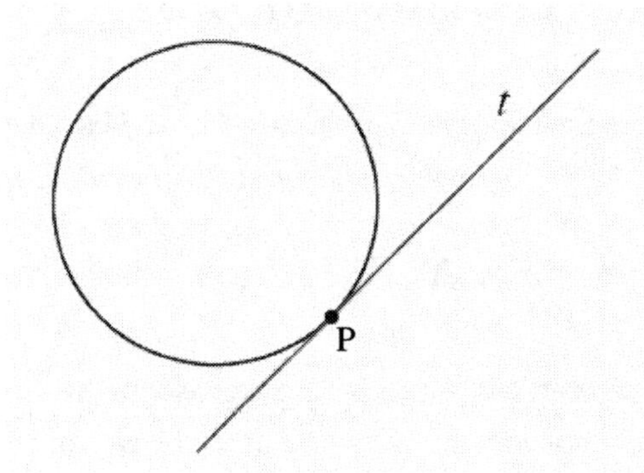

Figure 14.9
Tangent to a curve

Line t intersects the circle once and once only. It is a tangent to the curve at point P. However, in the more general case, lines intersecting curves at one position do not have the same slope as the curve at the point of intersection. An example can be seen in the next figure.

A secant line of a curve is a line that intersects at two or more points on the curve. The word *secant* comes from the Latin *secare, to cut.*

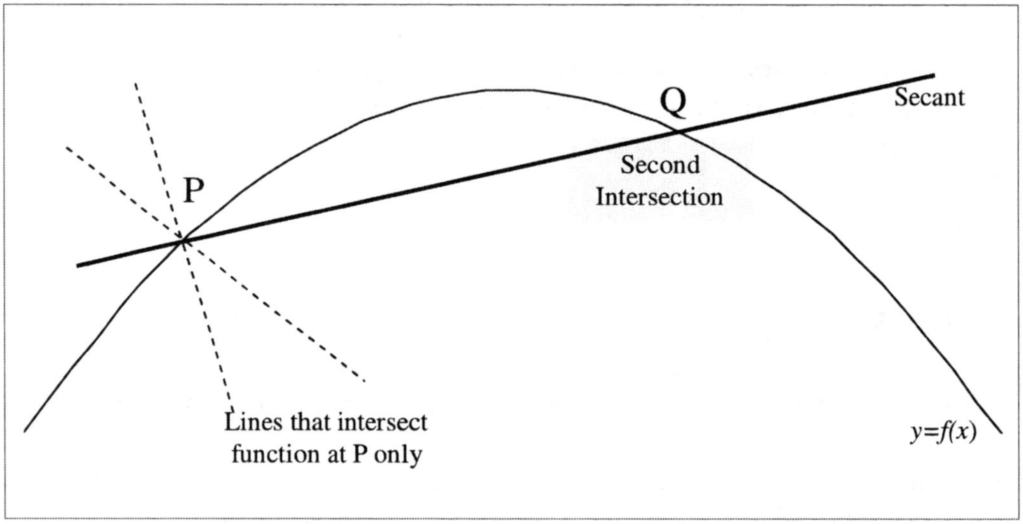

Figure 14.10
A secant line to a curve

The dashed lines intersect the curve once and once only but do not have the same slope as the curve at the point of intersection, P. They intersect only once because this is a line, and not a circle. The secant provides a basis for calculation of the tangent and therefore the gradient of the curve $y=f(x)$

Calculation of Gradients

We can get an approximation to the gradient of a curve at a particular point P by taking the secant line that intersects at P.

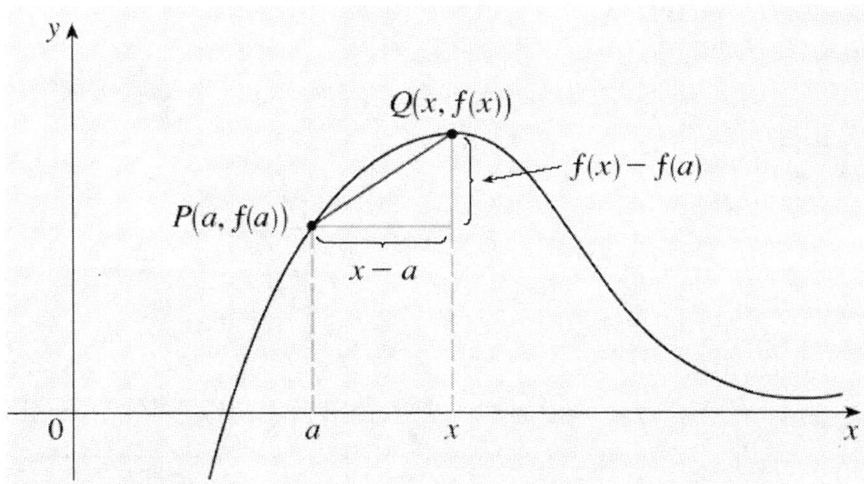

Figure 14.11
Taking a secant line of a curve

The tangent to the curve $y = f(x)$ at the point P(a, f(a)) is the line through P with slope

$$m = \lim_{x \to a} \frac{f(x) - f(a)}{x - a}$$

This is simply using the equation $$gradient = \frac{difference\ between\ y\ coordinates}{difference\ between\ x\ coordinates}$$

The y coordinates are f(x) and f(a), the x coordinates are x and a. In front of this is the term $\lim_{x \to a}$ This term refers to the gradient as the value of x approaches the value of a.

Now this curve can be treated in the same way as the straight line has been done previously, by renaming x as $a + h$, where h is the distance between a and x. If you take a situation where the distance h becomes smaller and smaller, x would get closer and closer to a, and so point Q on the curve would move closer to point P. If this happens, the secant, or cord that cuts the curve at two points, will cut across a smaller and smaller section of the curve, but will still be on the inside of the curve.

As point Q approaches point P, the line that was a secant on the inside of the curve, becomes a tangent on the outside of the curve. The distance h gets smaller and smaller, and the point when the secant becomes a tangent is when the distance approaches zero, referred to as $\lim_{h \to 0}$.

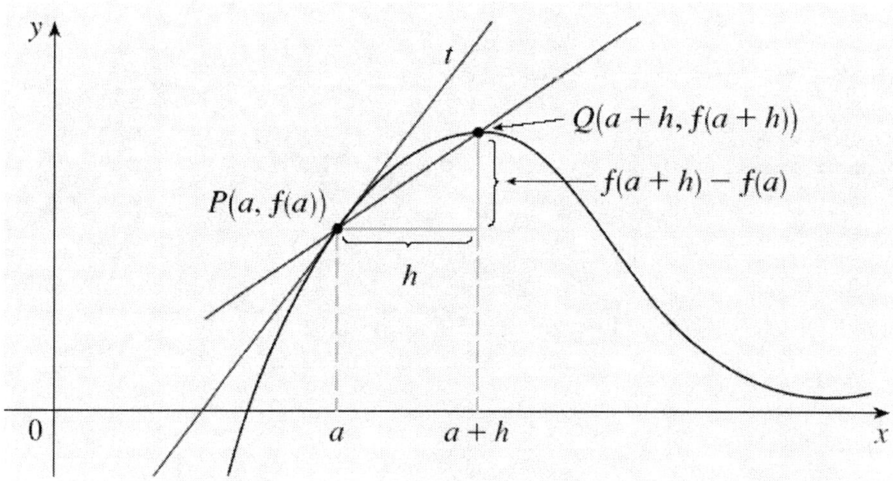

Figure 14.12
As a secant becomes a tangent

As with the similar treatment of a straight line, substituting $a + h$ for x gives

$$m = \lim_{h \to 0} \frac{f(a+h) - f(a)}{h}$$

The key difference with this equation and the one for a straight line is that this equation needs to be defined as one where h approaches zero, because it refers to a tangent of a curve.

The difference between the treatment of a straight line and a curve can be seen if you look at the difference between average and instantaneous velocities.

Average Velocity

Suppose an object moves along a straight line according to an equation of motion

$$s = f(t)$$

where s is the displacement of the object from the origin at time, t.
The function, f, that describes the motion is called the *position function* of the object.

Using the same process as was used for finding the gradient of a straight line, in the time interval from where $t = a$ to where $t = a + h$, the change in position is $f(a+h) - f(a)$. The average velocity over this time interval is

$$\text{average velocity} = \frac{\text{displacement}}{\text{time}} = \frac{f(a+h) - f(a)}{h}$$

This is the equation for average velocity.

Average and Instantaneous Velocities

If velocity varies with time, to find the velocity at any particular time it is necessary to draw a tangent to the curve:

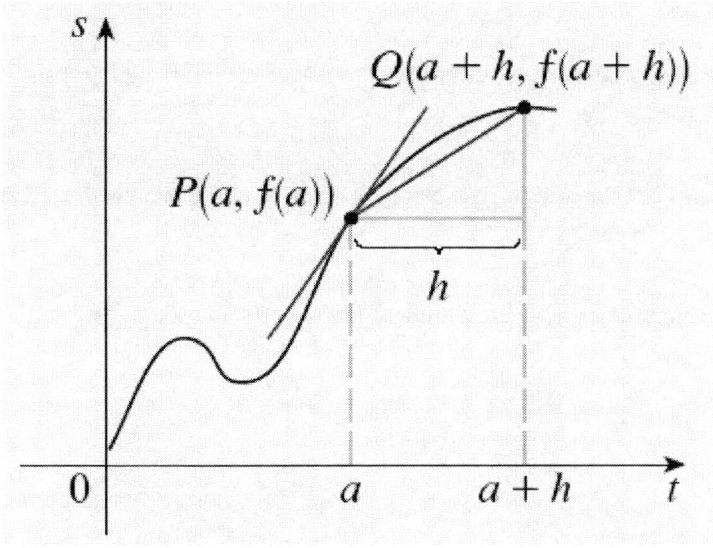

Figure 14.13 determining instantaneous velocities

Average velocity between points P and Q would be $m_{PQ} = \lim_{h \to 0} \frac{f(a+h) - f(a)}{h}$

What happens if we compute the average velocities over shorter and shorter time intervals $[a, a+h]$, in other words we let $h \to 0$? We now define the velocity $v(a)$ at time $t = a$ to be the limit of these average velocities:

$$v(a) = \lim_{h \to 0} \frac{f(a+h) - f(a)}{h}$$

This means that the velocity at time $t = a$ (also called the *instantaneous* velocity) is equal to the slope of the tangent line at P.

The Difference Quotient

Take the equation

$$gradient = \frac{difference\ between\ y\ coordinates}{difference\ between\ x\ coordinates}$$

This is actually saying that the gradient is a change in y over the change in x, or $\frac{\Delta y}{\Delta x}$.

This can also be written in the following way, referred to as the difference quotient.

$$\frac{\Delta y}{\Delta x} = \frac{f(x_2) - f(x_1)}{x_2 - x_1}$$

It is called the *average* rate of change of y with respect to x over the interval $[x_1,x_2]$ and can be interpreted as the slope (gradient) of the secant line PQ.

As x_2 approaches x_1, $\Delta x \to 0$ and the limit of the average rate of change becomes the (*instantaneous*) rate of change of y with respect to x at $x = x_1$.

This is interpreted as the slope of the tangent to the curve $y = f(x)$ at $P(x_1,f(x_1))$

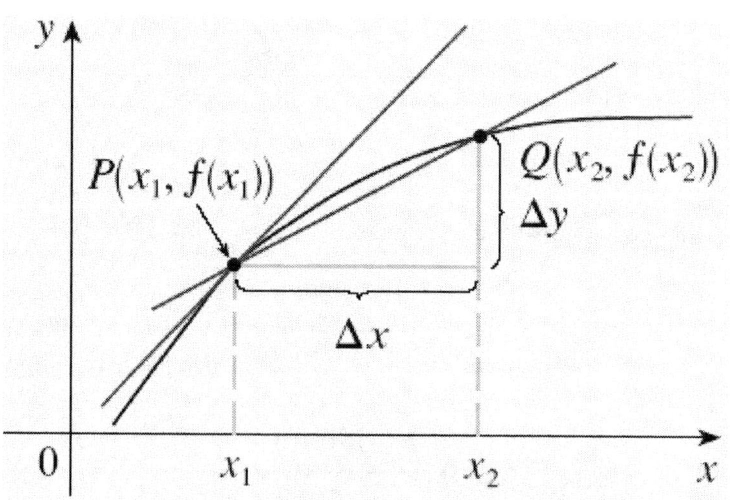

Figure 14.14
The difference quotient

$$\text{Instantaneous rate of change} = \lim_{\Delta x \to 0} \frac{\Delta y}{\Delta x} = \lim_{x_2 \to x_1} \frac{f(x_2) - f(x_1)}{x_2 - x_1}$$

This means that the tangent to the curve at point x_1 becomes equal to the slope of the line joining x_1 and x_2 when the distance between these two points (Δx) approaches zero.

This is where, for example, the value of x_2 approaches the value of x_1 and can be calculated from the difference in values for y over the difference in values for x of the slope of the secant, or tangent, to the curve.

Average of change = slope of secant PQ

Instantaneous rate of change = slope of tangent at P

The instantaneous rate of change is physically measurable and is the rate at a particular moment in time. The average rate of change over a particular time period is also a measurable quantity (think average speed cameras!).

The Derivative as a Function

If we go back to this equation

$$m_{PQ} = \lim_{h \to 0} \frac{f(a+h) - f(a)}{h}$$

and we now let the number a vary, replacing it by the variable x, we obtain a new function called f'. This is called the derivative of f and is defined as

$$f'(x) = \lim_{h \to 0} \frac{f(x+h) - f(x)}{h}$$

The value of the derivative, f' at x, referred to as $f'(x)$ can be interpreted as the slope (gradient) of the graph of f at the point $(x, f(x))$. It is the instantaneous rate of change of $y = f(x)$ at x.

The function f' is called the derivative of f because it has been 'derived' from f by the limiting operation above.

Maxima and Minima of a Function

Differentiation is frequently used to find the maxima and minima of a function. You have already seen this done in relation to a quadratic graph, using the Power Rule, in the introduction to this chapter. Here this process is explained in terms of calculus, and how it can relate to any curve that has maximum and minimum points.

The slopes of the tangents to the curve at positions A, B and C are all zero (Fig 14.15). As the value of x changes, the value of y remains the same. So while there is a change in the value of x, the change in y is equal to zero. This means that $\frac{\Delta y}{\Delta x} = 0$

so $f'(x) = 0.$

These are the stationary points of the function $y = f(x)$.

The values of these maximum and minimum points of the curve can be found using the Power Rule, but this does not tell you whether the values are positive or negative.

Figure 14.15 Showing maxima and minima of curve

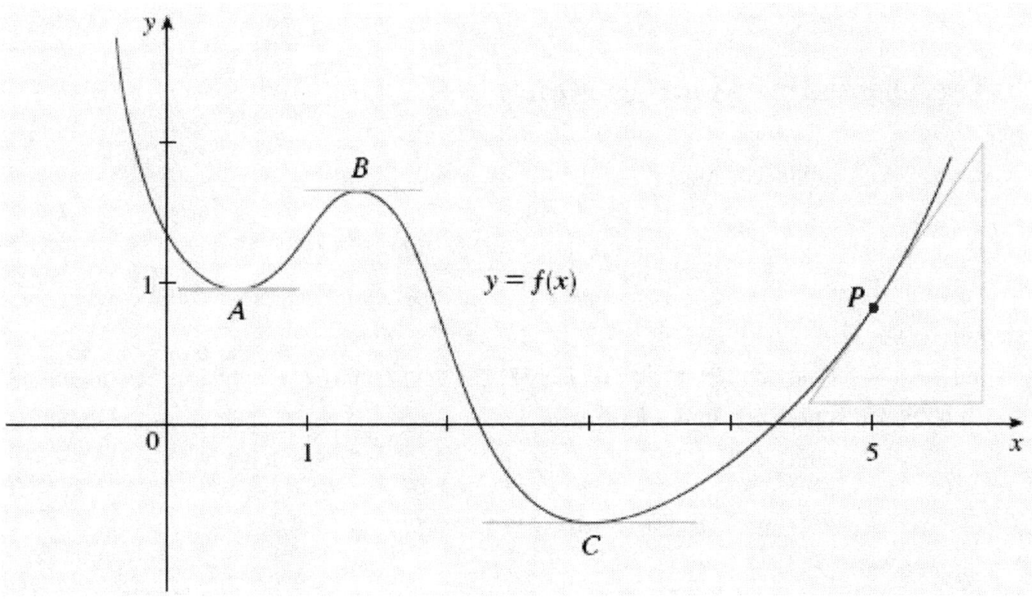

A second round of differentiation, using the Power Rule a second time, calculates what is referred to as the second derivitive, or $f''(x)$. The second derivatives of the function can be used to classify these as maxima (f'' negative) or minima (f'' positive).

The Power Rule

We are now ready to begin to look at some of the simple rules that will allow us to work out the derivatives of simple functions, $y = f(x)$. The first of these is the Power Rule.

The Power Rule is a method for differentiating expressions involving powers of a variable, *x*. This Rule states that the derivative of the function

$$f(x) = x^n$$

is

$$\frac{df(x)}{dx} = nx^{n-1}$$

i.e.

$$f'(x) = nx^{n-1}$$

Refer to the introduction to this chapter to get a more detailed description of how to use the Power Rule.

Proving the Power Rule

It is relatively straightforward to prove the Power Rule. Let's take a simple example by looking at differentiation of the function,

$$y = x^2$$

If the Power Rule is applied to this expression, you will get

$$\frac{dy}{dx} = 2x$$

But what does the $\frac{dy}{dx}$ mean?

We can predict the effect of making a small change, δy, to *y* and a small change δx, to *x*.

$$y + \delta y = (x + \delta x)^2$$

Expanding the contents of the brackets gives

$$y + \delta y = x^2 + 2x\delta x + \delta x^2$$

Now removing the equality $y = x^2$ gives the expression for the small change in x and y, rather than values of x and y.

$$\delta y = 2x\delta x + \delta x^2$$

To get the expression for a change in y over a change in x on one side of the equation, both sides of the equation are divided by δx

$$\frac{\delta y}{\delta x} = \frac{2x\delta x}{\delta x} + \frac{\delta x^2}{\delta x}$$

and simplified

$$\frac{\delta y}{\delta x} = 2x + \delta x$$

Finally, to derive the *instantaneous* rate of change of y with respect to x, we need to make the small change in x, δx vanishingly small i.e. take the limit $\delta x \rightarrow 0$. Under these circumstances, the average rate of change $\delta y/\delta x$ becomes the instantaneous rate and the way it is written changes to dy/dx to denote this. This is the derivative which we are seeking

$$\frac{dy}{dx} = 2x$$

More Useful Derivatives

What if we are faced with the problem of differentiating the expression?

$$y = \frac{1}{x}$$

The power rule can be applied in this situation. Note first that the following is an alternative way of writing the function in question

$$y = 1x^{-1}$$

You can then apply the Power Rule. The Power Rule can be written as an equation, rather than an expression

$$\frac{dy}{dx} = anx^{n-1}$$

$$\frac{dy}{dx} = 1 \times -1x^{-1-1} = -1x^{-2} = -\frac{1}{x^2}$$

The derivative of e^x is a special case. This is the only function which is equal to its own derivative.

Function $y = e^x$ Derivative $\frac{dy}{dx} = \frac{de^x}{dx} = e^x$

So the derivative of $y = e^{3x}$ is $\frac{dy}{dx} = 3e^{3x}$

and the derivative of $y = 10e^{3x}$ is $\frac{dy}{dx} = 30e^{3x}$

Note that while the Power Rule is applied to the equation as for previous examples, with the number in front of *e* being multiplied by the number in front of *x*, the power to which *e* is raised does not change.

The terms *x* and *y* apply to what would be plotted on each axis of a graph, and often you will see *x* represented as time, or *t*, so that differentiating

$y = e^t$ will give you $\frac{dy}{dt} = e^t$

and differentiating $y = e^{3t}$ will give you $\frac{dy}{dt} = 3e^{3t}$

You have looked at population growth with the equation

$$N = N_o e^{kt}$$

This equation looks at how a population increases with time. Calculus can be used to determine the instantaneous increase in population size. Remember that:

N= the number in the population

t = time

N_0 = the number in the population when $t = 0$

k = growth or decay constant

e = 2.718….

You can differentiate the equation above using the Power Rule as it applies to equations that include *e*. N_0, the term in front of *e*, is multiplied by the term in front of t, in the power to which e is raised, which is *k*. Note that here *t* is equivalent to *x* in the examples at the top of the page. This puts the term kN_0 in front of e. The power to which *e* is raised does not alter, as in the examples above.

So differentiating $N = N_o e^{kt}$ gives you $\frac{dN}{dt} = kN_o e^{kt}$.

Now you already know that $N = N_o e^{kt}$ and in the above equation the $N_o e^{kt}$ expression appears after k, and this expression is equal to N. Because of this, N can be substituted for $N_o e^{kt}$ and this gives you:

$$\frac{dN}{dt} = kN$$

Here the instantaneous rate of growth of a population is dependent on the growth rate and the number in the population. The term k can be taken to be the number of births, b, minus the number of deaths, d,

$$\frac{dN}{dt} = (b - d)N$$

so if there are more births than deaths, k would be positive and the population would be increasing so, in the infinitesimally small time period, the population would be larger at the end of the time period than it was at the beginning.
If k were negative, this means that the number of deaths exceeds the number of births, and so the population would be declining. If the births and deaths were equal, k would be equal to zero, so $\frac{dN}{dt} = 0$ and the tangent to the growth curve at this point would be horizontal.

In the above examples the letter in front of N is k, but there are many uses of this equation where a different letter is substituted for k. The equation itself is exactly the same, but the different letters used are described in different ways, for different purposes.

For example, you might come across this equation written as $\frac{dN}{dt} = \mu N$. Here μ is substituted for k. The term μ is described as the growth rate of a population,

Another example is where r is substituted for k, $\frac{dN}{dt} = rN$. Here r is described as the intrinsic rate of population growth.

This equation forms the base from which other parts are added to more accurately describe how a population grows. You might, for example, see the equation:

$$\frac{dN}{dt} = rN\left(\frac{K - N}{K}\right)$$

Here K refers to the carrying capacity, or the average number of individuals an environment can support; the resultant graph has an 'S-shaped' curve. So changes made to the equation result in changes to the growth curve, which reflect the way the population is growing.

Differentiating More Complex Functions

For any functions $f(x)$ and $g(x)$ and any real numbers a and b, the following rules apply to differentiation
The **constant multiple rule**

If $$h(x) = af(x)$$

then $$\frac{dh(x)}{dx} = a\frac{df(x)}{dx}$$

$$h'(x) = af'(x)$$

This can be written as $$(af)' = af'$$

Simple rules also govern the differentiation of functions formed from the addition and subtraction of other functions

The **addition rule** $$(f + g)' = f' + g'$$

The **subtraction rule** $$(f - g)' = f' - g'$$

These simple rules allow us to deal with more complicated examples
For example, the derivative of the function

$$h(x) = af(x) + bg(x)$$

with respect to x is $$\frac{dh(x)}{dx} = a\frac{df(x)}{dx} + b\frac{dg(x)}{dx}$$

often written as $$h'(x) = af'(x) + bg'(x)$$

or, more simply, $$h' = af' + bg'$$

Test your understanding

12. What is the difference between a secant and a tangent?

13. What is the difference between the following three terms?

$$\frac{\Delta y}{\Delta x} \qquad \frac{\delta y}{\delta x} \qquad \frac{dy}{dx}$$

Write down the derivatives of the following functions:

14. $y = x^3$

15. $y = \dfrac{1}{x^2}$

16. $y = -\dfrac{1}{x}$

17. $y = x^3 + x^2$

18. $y = x^4 - 2x^2 + \dfrac{2}{x^3}$

19. $y = \dfrac{1}{x^3} - \dfrac{2}{x} - 4$

20. $y = \dfrac{x^2}{4} + x^2$

Appendix 1
Solutions to Selected Exercises

from *Foundation Maths* Fourth Edition
Anthony Croft and Robert Davison

Solutions

Chapter 1

Self-assessment questions 1.1

1. BODMAS is a priority rule used when evaluating expressions: Brackets (do first), Of, Division, Multiplication (do secondly), Addition, Subtraction (do thirdly).
2. False. For example $(12 - 4) - 3$ is not the same as $12 - (4 - 3)$. The former is equal to 5 whereas the latter is equal to 11. The position of the brackets is clearly important.

Exercise 1.1

1. (a) $6 - 2 \times 2 = 6 - 4 = 2$
   (b) $(6 - 2) \times 2 = 4 \times 2 = 8$
   (c) $6 \div 2 - 2 = 3 - 2 = 1$
   (d) $(6 \div 2) - 2 = 3 - 2 = 1$
   (e) $6 - 2 + 3 \times 2 = 6 - 2 + 6 = 10$
   (f) $6 - (2 + 3) \times 2 = 6 - 5 \times 2$
   $= 6 - 10 = -4$
   (g) $(6 - 2) + 3 \times 2 = 4 + 3 \times 2 = 4 + 6$
   $= 10$
   (h) $\frac{16}{-2} = -8$ (i) $\frac{-24}{-3} = 8$
   (j) $(-6) \times (-2) = 12$
   (k) $(-2)(-3)(-4) = -24$
2. (a) $6 \times (12 - 3) + 1 = 6 \times 9 + 1$
   $= 54 + 1 = 55$
   (b) $6 \times 12 - (3 + 1) = 72 - 4 = 68$
   (c) $6 \times (12 - 3 + 1) = 6 \times 10 = 60$
   (d) $5 \times (4 - 3) + 2 = 5 \times 1 + 2 = 5 + 2 = 7$
   (e) $5 \times 4 - (3 + 2) = 20 - 5 = 15$
   or $5 \times (4 - 3 + 2) = 5 \times 3 = 15$
   (f) $5 \times (4 - (3 + 2)) = 5 \times (4 - 5)$
   $= 5 \times (-1) = -5$

Self-assessment questions 1.2

1. A prime number is a positive integer larger than 1 that cannot be expressed as the product of two smaller positive integers.
2. 2, 3, 5, 7, 11, 13, 17, 19, 23, 29
3. All even numbers have 2 as a factor and so can be expressed as the product of two smaller numbers. The exception to this is 2 itself, which can only be expressed as 1×2, and since these numbers are not both smaller than 2, then 2 is prime.

Exercise 1.2

1. 13, 2 and 29 are prime.
2. (a) 2×13 (b) $2 \times 2 \times 5 \times 5$
   (c) $3 \times 3 \times 3$ (d) 71
   (e) $2 \times 2 \times 2 \times 2 \times 2 \times 2$ (f) 3×29
   (g) 19×23 (h) 29×31
3. $30 = 2 \times 3 \times 5$
   $42 = 2 \times 3 \times 7$
   2 and 3 are common prime factors.

Chapter 2

Exercise 2.1

1. 16 cm = 160 mm = 0.16 m
2. 356 mm = 35.6 cm = 0.356 m
3. 156 m = 0.156 km
4. $A = 14 \times 7 = 98$ cm^2
5. $A = 17.5$ m^2, cost = £87.33
6. $A = 2 \times 8 = 16$ m^2
7. $s = (a + b + c)/2 = 17$,
   $A = \sqrt{s(s-a)(s-b)(s-c)}$
   $= \sqrt{2295} = 47.9$ cm^2
8. $A = \frac{1}{2}(a + b)h = 22$ cm^2
9. $A = \pi r^2 = 1017.9$ cm^2
10. $A = 78.5$ cm^2
11. $A = 26 = \pi r^2$ and so $r^2 = 26/\pi = 8.276$, then $r = 2.877$ cm
12. 13430 cm^2 = 1.3430 m^2
13. $V = \pi r^2 h = 1099.6$ cm^3 = 1.0996 litres
14. $V = \frac{4}{3}\pi r^3 = 2145$ cm^3
15. $86 = \frac{4}{3}\pi r^3$, therefore
    $r^3 = (3 \times 86)/(4 \times \pi) = 20.53$, then
    $r = 2.74$ cm
16. $A = \pi r^2 \times \theta/360 = 39.27$ cm^2
17. $V = \pi r^2 h = 28.274$ m^3, now
    1000 litres = 1 m^3
    and so 28.274 m^3 = 28 274 litres

Exercise 2.2

1 (a) 12 kg = 12 000 g
  (b) 12 kg = 12 000 000 mg
2. 168 mg = 0.168 g = 0.000168 kg or 1.68×10^{-4} kg
3. 0.005 tonnes = 5 kg
4. 875 tonnes = 875 000 kg = 8.75×10^5 kg
5. 3500 kg = 3.5 tonnes

Chapter 4

Self-assessment questions 4.1

1. (a) A fraction is formed by dividing a whole number by another whole number, for example, $\frac{11}{3}$.
   (b) If the top number (the numerator) of a fraction is greater than or equal to the bottom number (the denominator) then the fraction is improper. For example, $\frac{101}{100}$ and $\frac{9}{9}$ are both improper fractions.
   (c) If the top number is less than the bottom number then the fraction is proper. For example, $\frac{99}{100}$ is a proper fraction.
2. (a) The numerator is the 'top number' of a fraction.
   (b) The denominator is the 'bottom number' of a fraction. For example, in $\frac{17}{14}$, the numerator is 17 and the denominator is 14.

Exercise 4.1

1. (a) Proper (b) Proper (c) Improper
   (d) Proper (e) Improper

Self-assessment questions 4.2

1. True. For example $7 = \frac{7}{1}$.
2. The numerator and denominator can both be divided by their h.c.f. This simplifies the fraction.
3. $\frac{3}{4}$, $\frac{9}{12}$, $\frac{75}{100}$.

Exercise 4.2

1. (a) $\frac{18}{27}=\frac{2}{3}$ (b) $\frac{12}{20}=\frac{3}{5}$ (c) $\frac{15}{45}=\frac{1}{3}$
   (d) $\frac{25}{80}=\frac{5}{16}$ (e) $\frac{15}{60}=\frac{1}{4}$ (f) $\frac{90}{200}=\frac{9}{20}$
   (g) $\frac{15}{20}=\frac{3}{4}$ (h) $\frac{2}{18}=\frac{1}{9}$ (i) $\frac{16}{24}=\frac{2}{3}$
   (j) $\frac{30}{65}=\frac{6}{13}$ (k) $\frac{12}{21}=\frac{4}{7}$ (l) $\frac{100}{45}=\frac{20}{9}$
   (m) $\frac{6}{9}=\frac{2}{3}$ (n) $\frac{12}{16}=\frac{3}{4}$ (o) $\frac{13}{42}$
   (p) $\frac{13}{39}=\frac{1}{3}$ (q) $\frac{11}{33}=\frac{1}{3}$ (r) $\frac{14}{30}=\frac{7}{15}$
   (s) $-\frac{12}{16}=-\frac{3}{4}$ (t) $\frac{11}{-33}=-\frac{1}{3}$
   (u) $\frac{-14}{-30}=\frac{7}{15}$
2. $\frac{3}{4}=\frac{21}{28}$
3. $4=\frac{20}{5}$
4. $\frac{5}{12}=\frac{15}{36}$
5. $2=\frac{8}{4}$
6. $6=\frac{18}{3}$
7. $\frac{2}{3}=\frac{8}{12}, \frac{5}{4}=\frac{15}{12}, \frac{5}{6}=\frac{10}{12}$
8. $\frac{4}{9}=\frac{8}{18}, \frac{1}{2}=\frac{9}{18}, \frac{5}{6}=\frac{15}{18}$
9. (a) $\frac{1}{2}=\frac{6}{12}$ (b) $\frac{3}{4}=\frac{9}{12}$ (c) $\frac{5}{2}=\frac{30}{12}$
   (d) $5=\frac{60}{12}$ (e) $4=\frac{48}{12}$ (f) $12=\frac{144}{12}$

Self-assessment questions 4.3

1. The l.c.m. of the denominators is found. Each fraction is expressed in an equivalent form with this l.c.m. as denominator. Addition and subtraction can then take place.

Exercise 4.3

1. (a) $\frac{1}{4}+\frac{2}{3}=\frac{3}{12}+\frac{8}{12}=\frac{11}{12}$
   (b) $\frac{3}{5}+\frac{5}{3}=\frac{9}{15}+\frac{25}{15}=\frac{34}{15}$
   (c) $\frac{12}{14}-\frac{2}{7}=\frac{6}{7}-\frac{2}{7}=\frac{4}{7}$
   (d) $\frac{3}{7}-\frac{1}{2}+\frac{2}{21}=\frac{18}{42}-\frac{21}{42}+\frac{4}{42}=\frac{1}{42}$
   (e) $1\frac{1}{2}+\frac{4}{9}=\frac{3}{2}+\frac{4}{9}=\frac{27}{18}+\frac{8}{18}=\frac{35}{18}$
   (f) $2\frac{1}{4}-1\frac{1}{3}+\frac{1}{2}=\frac{9}{4}-\frac{4}{3}+\frac{1}{2}$
   $=\frac{27}{12}-\frac{16}{12}+\frac{6}{12}=\frac{17}{12}$
   (g) $\frac{10}{15}-1\frac{2}{5}+\frac{8}{3}=\frac{10}{15}-\frac{7}{5}+\frac{8}{3}$
   $=\frac{10}{15}-\frac{21}{15}+\frac{40}{15}=\frac{29}{15}$
   (h) $\frac{9}{10}-\frac{7}{16}+\frac{1}{2}-\frac{2}{5}=\frac{72}{80}-\frac{35}{80}+\frac{40}{80}-\frac{32}{80}$
   $=\frac{45}{80}=\frac{9}{16}$
2. (a) $\frac{7}{8}+\frac{1}{3}=\frac{21}{24}+\frac{8}{24}=\frac{29}{24}$
   (b) $\frac{1}{2}-\frac{3}{4}=\frac{2}{4}-\frac{3}{4}=-\frac{1}{4}$
   (c) $\frac{3}{5}+\frac{2}{3}+\frac{1}{2}=\frac{18}{30}+\frac{20}{30}+\frac{15}{30}=\frac{53}{30}=1\frac{23}{30}$
   (d) $\frac{3}{8}+\frac{1}{3}+\frac{1}{4}=\frac{9}{24}+\frac{8}{24}+\frac{6}{24}=\frac{23}{24}$

(e) $\frac{2}{3}-\frac{4}{7}=\frac{14}{21}-\frac{12}{21}=\frac{2}{21}$

(f) $\frac{1}{11}-\frac{1}{2}=\frac{2}{22}-\frac{11}{22}=-\frac{9}{22}$

(g) $\frac{3}{11}-\frac{5}{8}=\frac{24}{88}-\frac{55}{88}=\frac{-31}{88}$

3. (a) $\frac{5}{2}$ (b) $\frac{11}{3}$ (c) $\frac{41}{4}$ (d) $\frac{37}{7}$ (e) $\frac{56}{9}$

(f) $\frac{34}{3}$ (g) $\frac{31}{2}$ (h) $\frac{55}{4}$ (i) $\frac{133}{11}$ (j) $\frac{41}{3}$

(k) $\frac{113}{2}$

4. (a) $3\frac{1}{3}$ (b) $3\frac{1}{2}$ (c) $3\frac{3}{4}$ (d) $4\frac{1}{6}$

Self-assessment questions 4.4

1. The numerators are multiplied together to form the numerator of the product. The denominators are multiplied to form the denominator of the product.

Exercise 4.4

1. (a) $\frac{2}{3}\times\frac{6}{7}=\frac{2}{1}\times\frac{2}{7}=\frac{4}{7}$

(b) $\frac{8}{15}\times\frac{25}{32}=\frac{1}{15}\times\frac{25}{4}=\frac{1}{3}\times\frac{5}{4}=\frac{5}{12}$

(c) $\frac{1}{4}\times\frac{8}{9}=\frac{1}{1}\times\frac{2}{9}=\frac{2}{9}$

(d) $\frac{16}{17}\times\frac{34}{48}=\frac{1}{17}\times\frac{34}{3}=\frac{1}{1}\times\frac{2}{3}=\frac{2}{3}$

(e) $2\times\frac{3}{5}\times\frac{5}{12}=\frac{3}{5}\times\frac{5}{6}=\frac{3}{1}\times\frac{1}{6}=\frac{1}{2}$

(f) $2\frac{1}{3}\times1\frac{1}{4}=\frac{7}{3}\times\frac{5}{4}=\frac{35}{12}$

(g) $1\frac{3}{4}\times2\frac{1}{2}=\frac{7}{4}\times\frac{5}{2}=\frac{35}{8}$

(h) $\frac{3}{4}\times1\frac{1}{2}\times3\frac{1}{2}=\frac{3}{4}\times\frac{3}{2}\times\frac{7}{2}=\frac{63}{16}$

2. (a) $\frac{2}{3}\times\frac{3}{4}=\frac{1}{2}$ (b) $\frac{4}{7}\times\frac{21}{30}=\frac{6}{15}=\frac{2}{5}$

(c) $\frac{9}{10}\times80=72$ (d) $\frac{6}{7}\times42=36$

3. Yes, because $\frac{3}{4}\times\frac{12}{15}=\frac{12}{15}\times\frac{3}{4}$

4. (a) $-\frac{5}{21}$ (b) $-\frac{3}{8}$ (c) $-\frac{5}{11}$ (d) $\frac{10}{7}$

5. (a) $5\frac{1}{2}\times\frac{1}{2}=\frac{11}{2}\times\frac{1}{2}=\frac{11}{4}$

(b) $3\frac{3}{4}\times\frac{1}{3}=\frac{15}{4}\times\frac{1}{3}=\frac{5}{4}$

(c) $\frac{2}{3}\times5\frac{1}{9}=\frac{2}{3}\times\frac{46}{9}=\frac{92}{27}$

(d) $\frac{3}{4}\times11\frac{1}{2}=\frac{3}{4}\times\frac{23}{2}=\frac{69}{8}$

6. (a) $\frac{3}{5}\times11\frac{1}{4}=\frac{3}{5}\times\frac{45}{4}=\frac{27}{4}$

(b) $\frac{2}{3}\times15\frac{1}{2}=\frac{2}{3}\times\frac{31}{2}=\frac{31}{3}$

(c) $\frac{1}{4}\times\left(-8\frac{1}{3}\right)=\frac{1}{4}\times\left(-\frac{25}{3}\right)=-\frac{25}{12}$

Self-assessment questions 4.5

1. To divide one fraction by a second fraction, the second fraction is inverted (i.e. the numerator and denominator are interchanged) and then multiplication is performed.

Exercise 4.5

1. (a) $\frac{3}{4}\div\frac{1}{8}=\frac{3}{4}\times\frac{8}{1}=\frac{3}{1}\times\frac{2}{1}=6$

(b) $\frac{8}{9} \div \frac{4}{3} = \frac{8}{9} \times \frac{3}{4} = \frac{2}{9} \times \frac{3}{1} = \frac{2}{3}$

(c) $-\frac{2}{7} \div \frac{4}{21} = -\frac{2}{7} \times \frac{21}{4} = -\frac{2}{1} \times \frac{3}{4} = -\frac{3}{2}$

(d) $\frac{9}{4} \div 1\frac{1}{2} = \frac{9}{4} \div \frac{3}{2} = \frac{9}{4} \times \frac{2}{3} = \frac{3}{4} \times \frac{2}{1} = \frac{3}{2}$

(e) $\frac{5}{6} \div \frac{5}{12} = \frac{5}{6} \times \frac{12}{5} = \frac{1}{6} \times \frac{12}{1} = 2$

(f) $\frac{99}{100} \div 1\frac{4}{5} = \frac{99}{100} \div \frac{9}{5} = \frac{99}{100} \times \frac{5}{9}$

$= \frac{11}{100} \times \frac{5}{1} = \frac{11}{20}$

(g) $3\frac{1}{4} \div 1\frac{1}{8} = \frac{13}{4} \div \frac{9}{8} = \frac{13}{4} \times \frac{8}{9}$

$= \frac{13}{1} \times \frac{2}{9} = \frac{26}{9}$

(h) $\left(2\frac{1}{4} \div \frac{3}{4}\right) \times 2 = \left(\frac{9}{4} \times \frac{4}{3}\right) \times 2$

$= \left(\frac{3}{4} \times 4\right) \times 2 = 3 \times 2 = 6$

(i) $2\frac{1}{4} \div \left(\frac{3}{4} \times 2\right) = \frac{9}{4} \div \left(\frac{3}{4} \times \frac{2}{1}\right)$

$= \frac{9}{4} \div \frac{3}{2} = \frac{9}{4} \times \frac{2}{3} = \frac{3}{4} \times \frac{2}{1} = \frac{3}{2}$

(j) $6\frac{1}{4} \div 2\frac{1}{2} + 5 = \frac{25}{4} \div \frac{5}{2} + 5$

$= \frac{25}{4} \times \frac{2}{5} + 5 = \frac{5}{2} + 5 = \frac{15}{2}$

(k) $6\frac{1}{4} \div \left(2\frac{1}{2} + 5\right) = \frac{25}{4} \div \left(\frac{5}{2} + 5\right)$

$= \frac{25}{4} \div \frac{15}{2} = \frac{25}{4} \times \frac{2}{15} = \frac{5}{4} \times \frac{2}{3} = \frac{5}{6}$

Self-assessment questions 4.6

1. Largest is 23.01; smallest is 23.0
2. 0.1

Exercise 4.6

1. (a) $\frac{7}{10}$ (b) $\frac{4}{5}$ (c) $\frac{9}{10}$
2. (a) $\frac{11}{20}$ (b) $\frac{79}{500}$ (c) $\frac{49}{50}$ (d) $\frac{99}{1000}$
3. (a) $4\frac{3}{5}$ (b) $5\frac{1}{5}$ (c) $8\frac{1}{20}$ (d) $11\frac{59}{100}$
   (e) $121\frac{9}{100}$
4. (a) 0.697 (b) 0.083 (c) 0.517

Self-assessment questions 4.7

1. Writing a number to 2 (or 3 or 4 etc) significant figures is a way of approximatin the number. The number of s.f. is the maximum number of non-zero digits in the approximation. The approximation is as close as possible to the original number.
2. The digits after the decimal point are considered. To write to 1 d.p. we consider th first 2 digits; to write to 2 d.p. we consider th

first 3 digits, and so on. If the last digit considered is 5 or greater, we round up the previous digit; otherwise we round down.

Exercise 4.7

1. (a) 6960 (b) 70.4 (c) 0.0123 (d) 0.0110 (e) 45.6 (f) 2350
2. (a) 66.00 (b) 66.0 (c) 66 (d) 70 (e) 66.00 (f) 66.0
3. (a) 10 (b) 10.0
4. (a) 65.456 (b) 65.46 (c) 65.5 (d) 65.456 (e) 65.46 (f) 65.5 (g) 65 (h) 70

Chapter 5

Self-assessment questions 5.1

1. Converting fractions to percentages allows for easy comparison of numbers.

Exercise 5.1

1. 23% of $124 = \dfrac{23}{100} \times 124 = 28.52$
2. (a) $\dfrac{9}{11} = \dfrac{9}{11} \times 100\% = \dfrac{900}{11}\ \% = 81.82\%$

   (b) $\dfrac{15}{20} = \dfrac{15}{20} \times 100\% = 75\%$

   (c) $\dfrac{9}{10} = \dfrac{9}{10} \times 100\% = 90\%$

   (d) $\dfrac{45}{50} = \dfrac{45}{50} \times 100\% = 90\%$

   (e) $\dfrac{75}{90} = \dfrac{75}{90} \times 100\% = 83.33\%$
3. $\dfrac{13}{12} = \dfrac{13}{12} \times 100\% = 108.33\%$
4. 217% of $500 = \dfrac{217}{100} \times 500 = 1085$
5. New weekly wage is 106% of £400,

   106% of $400 = \dfrac{106}{100} \times 400 = 424$

   The new weekly wage is £424.
6. 17% of $1200 = \dfrac{17}{100} \times 1200 = 204$

   The debt is decreased by £204 to £1200 − 204 = £996.
7. (a) $50\% = \dfrac{50}{100} = 0.5$

   (b) $36\% = \dfrac{36}{100} = 0.36$

   (c) $75\% = \dfrac{75}{100} = 0.75$

   (d) $100\% = \dfrac{100}{100} = 1$

   (e) $12.5\% = \dfrac{12.5}{100} = 0.125$
8. £204.80
9. £1125

Self-assessment questions 5.2

1. True

Exercise 5.2

1. $8 + 1 + 3 = 12$. The first number is $\frac{8}{12}$ of 180, that is, 120; the second number is $\frac{1}{12}$ of 180, that is, 15; and the third number is $\frac{3}{12}$ of 180, that is, 45. Hence 180 is divided into 120, 15 and 45.

2. $1+1+3=5$. We calculate $\frac{1}{5}$ of 930 to be 186 and $\frac{3}{5}$ of 930 to be 558. The length is divided into 186 cm, 186 cm and 558 cm.
3. $2+3+4=9$. The first piece is $\frac{2}{9}$ of 6 m, that is, 1.33 m; the second piece is $\frac{3}{9}$ of 6 m, that is, 2 m; the third piece is $\frac{4}{9}$ of 6 m, that is, 2.67 m.
4. $1+2+3+4=10$: $\frac{1}{10}$ of $1200=120$; $\frac{2}{10}$ of $1200=240$; $\frac{3}{10}$ of $1200=360$; $\frac{4}{10}$ of 1200 $=480$. The number 1200 is divided into 120, 240, 360 and 480.
5. $2\frac{3}{4}:1\frac{1}{2}:2\frac{1}{4}=\frac{11}{4}:\frac{3}{2}:\frac{9}{4}=11:6:9$

Now, $11+6+9=26$, so

$$\frac{11}{26}\times 2600=1100 \qquad \frac{6}{26}\times 2600=600$$

$$\frac{9}{26}\times 2600=900$$

Alan receives £1100, Bill receives £600 and Claire receives £900.

6. 8 kg, 10.67 kg, 21.33 kg
7. (a) 1 : 2 (b) 1 : 2 (c) 1 : 2 : 4 (d) 1 : 21
8. 24, 84

Chapter 6

Self-assessment questions 6.1

1. 'Algebra' refers to the manipulation of symbols, as opposed to the manipulation of numbers.
2. The product of the two numbers is written as mn.
3. An algebraic fraction is formed by dividing one algebraic expression by another algebraic expression. The 'top' of the fraction is the numerator; the 'bottom' of the fraction is the denominator.
4. Superscripts and subscripts are located in different positions, relative to the symbol. Superscripts are placed high; subscripts are placed low.
5. A variable can have many different values; a constant has one, fixed value.

Self-assessment questions 6.2

1. In the expression a^x, a is the base and x is the power.
2. 'Index' is another word meaning 'power'.
3. $(xyz)^2$ means $(xyz)(xyz)$, which can be written as $x^2y^2z^2$. Clearly this is distinct from xyz^2, in which only the quantity z is squared.
4. $(-3)^4=(-3)(-3)(-3)(-3)=81$.

$-3^4=-(3)(3)(3)(3)=-81$. Here, the power (4) has the higher priority.

Exercise 6.1

1. $2^4=16$; $(\frac{1}{2})^2=\frac{1}{4}$; $1^8=1$; $3^5=243$; $0^3=0$.
2. $10^4=10000$; $10^5=100000$; $10^6=1000000$.
3. $11^4=14641$; $16^8=4294967296$; $39^4=2313441$; $1.5^7=17.0859375$.
4. (a) $a^4b^2c=a\times a\times a\times a\times b\times b\times c$
   (b) $xy^2z^4=x\times y\times y\times z\times z\times z\times z$
5. (a) x^4y^2 (b) $x^2y^2z^3$ (c) $x^2y^2z^2$
   (d) $a^2b^2c^2$
6. (a) $7^4=2401$ (b) $7^5=16807$
   (c) $7^4\times 7^5=40353607$
   (d) $7^9=40353607$ (e) $8^3=512$
   (f) $8^7=2097152$
   (g) $8^3\times 8^7=1073741824$
   (h) $8^{10}=1073741824$
   The rule states that $a^m\times a^n=a^{m+n}$; the powers are added.
7. $(-3)^3=-27$; $(-2)^2=4$; $(-1)^7=-1$; $(-1)^4=1$.
8. -4492.125; 324; -0.03125.
9. (a) 36 (b) 9 (c) -64 (d) -8

$-6^2 = -36$, $-3^2 = -9$,
$-4^3 = -64$, $-2^3 = -8$

Self-assessment questions 6.3

1. An algebraic expression is any quantity comprising symbols and operations (i.e. +, −, ×, ÷): for example, x, $2x^2$ and $3x + 2y$ are all algebraic expressions. An algebraic formula relates two or more quantities and must contain an '=' sign. For example, $A = \pi r^2$, $V = \pi r^2 h$ and $S = ut + \frac{1}{2}at^2$ are all algebraic formulae.

Exercise 6.2

1. 60
2. 69
3. (a) 314.2 cm^2 (b) 28.28 cm^2 (c) 0.126 cm^2
4. $3x^2 = 3 \times 4^2 = 3 \times 16 = 48$; $(3x)^2 = 12^2 = 144$
5. $5x^2 = 5(-2)^2 = 20$; $(5x)^2 = (-10)^2 = 100$
6. (a) 33.95 (b) 23.5225 (c) 26.75 (d) 109.234125
7. (a) $a + b + c = 25.5$ (b) $ab = 46.08$ (c) $bc = 32.76$ (d) $abc = 419.328$
8. $C = \frac{5}{9}(100 - 32) = \frac{5}{9}(68) = 37.78$
9. (a) $x^2 = 49$ (b) $-x^2 = -49$ (c) $(-x)^2 = (-7)^2 = 49$
10. (a) 4 (b) 4 (c) −4 (d) 12 (e) −12 (f) 36
11. (a) 3 (b) 9 (c) −1 (d) 36 (e) −36 (f) 144
12. $x^2 - 7x + 2 = (-9)^2 - 7(-9) + 2 = 81 + 63 + 2 = 146$
13. $2x^2 + 3x - 11 = 2(-3)^2 + 3(-3) - 11 = 18 - 9 - 11 = -2$
14. $-x^2 + 3x - 5 = -(-1)^2 + 3(-1) - 5 = -1 - 3 - 5 = -9$
15. 0
16. (a) 49 (b) 43 (c) 1 (d) 5
17. (a) 21 (b) 27 (c) 0 (d) $\frac{1}{6}$
18. (a) 3 (b) $3\frac{4}{5}$ (c) 23 (d) 23
19. (a) $-\frac{1}{2}$ (b) 4 (c) 32
20. (a) 0 (b) 16 (c) $40\frac{1}{2}$ (d) $2\frac{1}{2}$
21. (a) 17 (b) −0.5 (c) 7
22. (a) 6000 (b) 2812.5
23. (a) 17151 (b) 276951

Chapter 7

Self-assessment questions 7.1

1. $a^m \times a^n = a^{m+n}$
   $\frac{a^m}{a^n} = a^{m-n}$
   $(a^m)^n = a^{mn}$
2. a^0 is 1
3. x^1 is simply x.

Exercise 7.1

1. (a) $5^7 \times 5^{13} = 5^{20}$ (b) $9^8 \times 9^5 = 9^{13}$ (c) $11^2 \times 11^3 \times 11^4 = 11^9$
2. (a) $15^3/15^2 = 15^1 = 15$ (b) $4^{18}/4^9 = 4^9$ (c) $5^{20}/5^{19} = 5^1 = 5$
3. (a) a^{10} (b) a^9 (c) b^{22}
4. (a) x^{15} (b) y^{21}
5. $19^8 \times 17^8$ cannot be simplified using the laws of indices because the two bases are not the same.
6. (a) $(7^3)^2 = 7^{3\times 2} = 7^6$ (b) $(4^2)^8 = 4^{16}$ (c) $(7^9)^2 = 7^{18}$
7. $1/(5^3)^8 = 1/5^{24}$

8. (a) x^5y^5 (b) $a^3b^3c^3$
9. (a) $x^{10}y^{20}$ (b) $81x^6$ (c) $-27x^3$ (d) x^8y^{12}
10. (a) z^3 (b) y^2 (c) 1

Self-assessment questions 7.2

1. a^{-m} is the same as $\frac{1}{a^m}$. For example, 5^{-2} is the same as $\frac{1}{5^2}$.

Exercise 7.2

1. (a) $\frac{1}{4}$ (b) $\frac{1}{8}$ (c) $\frac{1}{9}$ (d) $\frac{1}{27}$ (e) $\frac{1}{25}$

   (f) $\frac{1}{16}$ (g) $\frac{1}{9}$ (h) $\frac{1}{121}$ (i) $\frac{1}{7}$
2. (a) 0.1 (b) 0.01 (c) 0.000001 (d) 0.01 (e) 0.001 (f) 0.0001
3. (a) $\frac{1}{x^4}$ (b) x^5 (c) $\frac{1}{x^7}$ (d) $\frac{1}{y^2}$

   (e) $y^1 = y$ (f) $\frac{1}{y^1} = \frac{1}{y}$ (g) $\frac{1}{y^2}$

   (h) $\frac{1}{z^1} = \frac{1}{z}$ (i) $z^1 = z$
4. (a) $x^{-3} = \frac{1}{x^3}$ (b) $x^{-5} = \frac{1}{x^5}$ (c) $x^{-1} = \frac{1}{x}$

   (d) x^5 (e) $x^{-13} = \frac{1}{x^{13}}$ (f) $x^{-8} = \frac{1}{x^8}$

   (g) $x^{-9} = \frac{1}{x^9}$ (h) $x^{-4} = \frac{1}{x^4}$
5. (a) a^{11} (b) x^{-16} (c) x^{-18} (d) 4^{-6}
6. (a) 0.001 (b) 0.0001 (c) 0.00001
7. $4^{-8}/4^{-6} = 4^{-2} = 1/4^2 = \frac{1}{16}$; $3^{-5}/3^{-8} = 3^3 = 27$

Self-assessment questions 7.3

1. $x^{\frac{1}{2}}$ is the square root of x. $x^{\frac{1}{3}}$ is the cube root of x.
2. 10, −10. Negative numbers, such as −100, do not have square roots, since squaring a number always gives a positive result.

Exercise 7.3

1. (a) $64^{1/3} = \sqrt[3]{64} = 4$ since $4^3 = 64$
   (b) $144^{1/2} = \sqrt{144} = \pm 12$
   (c) $16^{-1/4} = 1/16^{1/4} = 1/\sqrt[4]{16} = \pm\frac{1}{2}$
   (d) $25^{-1/2} = 1/25^{1/2} = 1/\sqrt{25} = \pm\frac{1}{5}$
   (e) $1/32^{-1/5} = 32^{1/5} = \sqrt[5]{32} = 2$ since $2^5 = 32$
2. (a) $(3^{-1/2})^4 = 3^{-2} = 1/3^2 = \frac{1}{9}$
   (b) $(8^{1/3})^{-1} = 8^{-1/3} = 1/8^{1/3} = 1/\sqrt[3]{8} = \frac{1}{2}$
3. (a) $8^{1/2}$ (b) $12^{1/3}$
   (c) $16^{1/4}$ (d) $13^{3/2}$ (e) $4^{7/3}$
4. (a) $x^{1/2}$ (b) $y^{1/3}$
   (c) $x^{5/2}$ (d) $5^{7/3}$

Exercise 7.4

1. (a) 743
   (b) 74300
   (c) 70
   (d) 0.0007
2. (a) 3×10^2
   (b) 3.56×10^2
   (c) 0.32×10^2
   (d) 0.0057×10^2

Self-assessment questions 7.4

1. Scientific notation is useful for writing very large or extremely small numbers in a concise way. It is easier to manipulate such numbers when they are written using scientific notation.

Exercise 7.5

1. (a) $45 = 4.5 \times 10^1$
   (b) $45000 = 4.5 \times 10^4$
   (c) $-450 = -4.5 \times 10^2$
   (d) $90000000 = 9.0 \times 10^7$
   (e) $0.15 = 1.5 \times 10^{-1}$
   (f) $0.00036 = 3.6 \times 10^{-4}$

(g) 3.5 is already in standard form.
(h) $-13.2 = -1.32 \times 10^1$
(i) $1000000 = 1 \times 10^6$
(j) $0.0975 = 9.75 \times 10^{-2}$
(k) $45.34 = 4.534 \times 10^1$

2. (a) $3.75 \times 10^2 = 375$
(b) $3.97 \times 10^1 = 39.7$
(c) $1.875 \times 10^{-1} = 0.1875$
(d) $-8.75 \times 10^{-3} = -0.00875$

3. (a) 2.4×10^8 (b) 7.968×10^8
(c) 1.044×10^{-4}
(d) 1.526×10^{-1}
(e) 5.293×10^2

Chapter 8

Exercise 8.1

1. (a) $-5p + 19q$ (b) $-5r - 13s + z$ (c) not possible to simplify
(d) $8x^2 + 3y^2 - 2y$ (e) $4x^2 - x + 9$
2. (a) $-12y + 8p + 9q$ (b) $21x^2 - 11x^3 + y^3$
(c) $7xy + y^2$ (d) $2xy$ (e) 0

Self-assessment questions 8.1

1. Positive
2. Negative

Exercise 8.2

1. (a) 84 (b) 84 (c) 84
2. (a) 40 (b) 40
3. (a) $14z$ (b) $30y$ (c) $6x$ (d) $27a$
(e) $55a$ (f) $6x$
4. (a) $20x^2$ (b) $6y^3$ (c) $22u^2$ (d) $8u^2$
(e) $26z^2$
5. (a) $21x^2$ (b) $21a^2$ (c) $14a^2$
6. (a) $15y^2$ (b) $8y$
7. (a) $a^3b^2c^2$ (b) x^3y^2 (c) x^2y^4
8. No difference; both equal x^2y^4.
9. $(xy^2)(xy^2) = x^2y^4$; $xy^2 + xy^2 = 2xy^2$
10. (a) $-21z^2$ (b) $-4z$
11. (a) $-3x^2$ (b) $2x$
12. (a) $2x^2$ (b) $-3x$

Exercise 8.3

1. (a) $4x + 4$ (b) $-4x - 4$ (c) $4x - 4$
(d) $-4x + 4$
2. (a) $5x - 5y$ (b) $19x + 57y$ (c) $8a + 8b$
(d) $5y + xy$ (e) $12x + 48$
(f) $17x - 153$ (g) $-a + 2b$ (h) $x + \frac{1}{2}$
(i) $6m - 12m^2 - 9mn$
3. (a) $18 - 13x - 26 = -8 - 13x$ (b) $x^2 + xy$ will not simplify any further.
4. (a) $x^2 + 7x + 6$ (b) $x^2 + 9x + 20$
(c) $x^2 + x - 6$ (d) $x^2 + 5x - 6$
(e) $xm + ym + nx + yn$
(f) $12 + 3y + 4x + yx$ (g) $25 - x^2$
(h) $51x^2 - 79x - 10$
5. (a) $x^2 - 4x - 21$ (b) $6x^2 + 11x - 7$
(c) $16x^2 - 1$ (d) $x^2 - 9$ (e) $6 + x - 2x^2$
6. (a) $\frac{29}{2}x - \frac{5}{2}y$ (b) $\frac{5}{4}x + \frac{5}{4}$
7. (a) $-x + y$ (b) $-a - 2b$ (c) $-\frac{3}{2}p - \frac{1}{2}q$
8. $(x + 1)(x + 2) = x^2 + 3x + 2$. So
$(x + 1)(x + 2)(x + 3) = (x^2 + 3x + 2)(x + 3)$
$= (x^2 + 3x + 2)(x) + (x^2 + 3x + 2)(3)$
$= x^3 + 3x^2 + 2x + 3x^2 + 9x + 6$
$= x^3 + 6x^2 + 11x + 6$

Chapter 9

Self-assessment questions 9.1

1. There are no factors common to both numerator and denominator and hence cancellation is not possible. In particular 3 is not a factor of the denominator.
2. Same as Q1 – there are no factors common to numerator and denominator. In particular x is a factor neither of the denominator nor of the numerator.
3. $$\frac{x+1}{2x+2} = \frac{x+1}{2(x+1)} = \frac{1}{2}$$

   The common factor, $x+1$, has been cancelled.

Exercise 9.1

1. (a) $\frac{3x}{y}$ (b) $\frac{9}{x}$ (c) $3y$ (d) $3x$ (e) 9 (f) 3
2. (a) $\frac{5x}{y}$ (b) $\frac{3x}{y}$ (c) $15y$ (d) 15 (e) $-x^2$ (f) $-\frac{1}{y^4} = -y^{-4}$ (g) $\frac{1}{y} = y^{-1}$ (h) y^{-7}
3. (a) $\frac{1}{3+2x}$ (b) $1+2x$ (c) $\frac{1}{2+7x}$ (d) $\frac{x}{2+7x}$ (e) $\frac{x}{1+7x}$ (f) $\frac{1}{7x+y}$ (g) $\frac{y}{7x+y}$ (h) $\frac{x}{7x+y}$
4. (a) $5x+1$ (b) $\frac{3(5x+1)}{3(x+2y)} = \frac{5x+1}{x+2y}$ (c) $\frac{3}{x+2}$ (d) $\frac{3}{y+2}$ (e) $\frac{13}{x+5}$ (f) $\frac{17}{9y+4}$
5. (a) $\frac{1}{3+2x}$ (b) $\frac{2}{x+7}$ (c) $\frac{2(x+4)}{(x-2)(x+4)} = \frac{2}{x-2}$ (d) $\frac{7}{ab+9}$ (e) $\frac{y}{y+1}$
6. (a) $\frac{1}{x-2}$ (b) $\frac{2(x-2)}{(x-2)(x+3)} = \frac{2}{x+3}$ (c) $\frac{1}{x+2}$ (d) $\frac{(x+1)(x+1)}{(x+1)(x-3)} = \frac{x+1}{x-3}$ (e) $\frac{2}{x-3}$ (f) $\frac{1}{x-3}$ (g) $\frac{1}{2(x-3)}$ (h) $\frac{2}{x-3}$ (i) $\frac{1}{2(x+4)}$ (j) $\frac{1}{2}$ (k) 2 (l) $x+4$ (m) $\frac{1}{x-3}$ (n) $\frac{1}{x+4}$ (o) $\frac{1}{2}$ (p) $\frac{x+4}{2x+9}$

Self-assessment questions 9.2

1. True

Exercise 9.2

1. (a) $\frac{y}{6}$ (b) $\frac{z}{6}$ (c) $\frac{2}{5y}$ (d) $\frac{2}{5x}$ (e) $\frac{3x}{4y}$ (f) $\frac{3x^2}{5y}$ (g) $\frac{3x}{5y}$ (h) $\frac{7x}{16y}$ (i) $\frac{1}{4x}$ (j) $\frac{x}{4}$ (k) $\frac{1}{x}$ (l) $\frac{x}{9}$ (m) $\frac{1}{x}$ (n) $\frac{1}{9x}$ (o) $\frac{x}{2}$
2. (a) $\frac{1}{x}$ (b) $\frac{x}{4}$ (c) 1 (d) x (e) $\frac{1}{x}$

(f) $\frac{4}{x}$ (g) $\frac{6}{x}$

3. (a) $\frac{a}{20}$ (b) $\frac{5a}{4b}$ (c) $\frac{1}{2ab}$ (d) $\frac{6x^2}{y^3}$
   (e) $\frac{3b}{5a^2}$ (f) $\frac{x}{4y}$ (g) $\frac{x}{3(x+y)}$ (h) $\frac{1}{3(x+4)}$
4. (a) $\frac{3y}{z^3}$ (b) $\frac{3+x}{y}$ (c) $\frac{x}{12}$ (d) $\frac{b}{c}$
5. (a) $\frac{1}{x+4}$ (b) $\frac{4(x-2)}{x}$
   (c) $\frac{12ab}{5ef} \times \frac{f}{4ab^2} = \frac{3}{5eb}$
   (d) $\frac{x+3y}{2x} \times \frac{4x^2}{y} = \frac{2x(x+3y)}{y}$ (e) $\frac{9}{xyz}$
6. $\frac{2}{x+3}$
7. $\frac{x+4}{x+3}$

Self-assessment questions 9.3

1. The l.c.d. is the simplest expression that is divisible by all the given denominators. To find the l.c.d. first factorise all denominators. The l.c.d. is then formed by including the minimum number of factors from the denominators such that each denominator can divide into the l.c.d.

Exercise 9.3

1. (a) $\frac{5z}{6}$ (b) $\frac{7x}{12}$ (c) $\frac{6y}{25}$
2. (a) $\frac{x+2}{2x}$ (b) $\frac{1+2x}{2}$ (c) $\frac{1+3y}{3}$
   (d) $\frac{y+3}{3y}$ (e) $\frac{8y+1}{y}$
3. (a) $\frac{10-x}{2x}$ (b) $\frac{5+2x}{x}$ (c) $\frac{9-x}{3x}$
   (d) $\frac{2x-3}{6}$ (e) $\frac{9+x}{3x}$
4. (a) $\frac{3}{x}+\frac{4}{y}=\frac{3y}{xy}+\frac{4x}{xy}=\frac{3y+4x}{xy}$
   (b) $\frac{3}{x^2}+\frac{4y}{x}=\frac{3}{x^2}+\frac{4xy}{x^2}=\frac{3+4xy}{x^2}$
   (c) $\frac{4ab}{x}+\frac{3ab}{2y}=\frac{8aby}{2xy}+\frac{3abx}{2xy}=\frac{8aby+3abx}{2xy}$
   (d) $\frac{4xy}{a}+\frac{3xy}{2b}=\frac{8xyb}{2ab}+\frac{3xya}{2ab}$
   $=\frac{8xyb+3xya}{2ab}$
   (e) $\frac{3}{x}-\frac{6}{2x}=\frac{3}{x}-\frac{3}{x}=0$
   (f) $\frac{3x}{2y}-\frac{7y}{4x}=\frac{6x^2}{4xy}-\frac{7y^2}{4xy}=\frac{6x^2-7y^2}{4xy}$
   (g) $\frac{3}{x+y}-\frac{2}{y}=\frac{3y}{(x+y)y}-\frac{2(x+y)}{(x+y)y}$
   $=\frac{3y-2x-2y}{(x+y)y}=\frac{y-2x}{(x+y)y}$
   (h) $\frac{1}{a+b}-\frac{1}{a-b}$
   $=\frac{a-b}{(a+b)(a-b)}-\frac{a+b}{(a+b)(a-b)}$
   $=\frac{a-b-a-b}{(a+b)(a-b)}=\frac{-2b}{(a+b)(a-b)}$
   (i) $2x+\frac{1}{2x}=\frac{4x^2}{2x}+\frac{1}{2x}=\frac{4x^2+1}{2x}$
   (j) $2x-\frac{1}{2x}=\frac{4x^2}{2x}-\frac{1}{2x}=\frac{4x^2-1}{2x}$
5. (a) $\frac{x}{y}+\frac{3x^2}{z}=\frac{xz}{yz}+\frac{3x^2y}{yz}=\frac{xz+3x^2y}{yz}$
   (b) $\frac{4}{a}+\frac{5}{b}=\frac{4b+5a}{ab}$
   (c) $\frac{6x}{y}-\frac{2y}{x}=\frac{6x^2-2y^2}{xy}$
   (d) $3x-\frac{3x+1}{4}=\frac{12x}{4}-\frac{3x+1}{4}$
   $=\frac{12x-3x-1}{4}=\frac{9x-1}{4}$

(e) $\frac{5a}{12}+\frac{9a}{18}=\frac{15a+18a}{36}=\frac{33a}{36}=\frac{11a}{12}$

(f) $\frac{x-3}{4}+\frac{3}{5}=\frac{5(x-3)+12}{20}$

$=\frac{5x-15+12}{20}=\frac{5x-3}{20}$

6. (a) $\frac{2x+3}{(x+1)(x+2)}$ (b) $\frac{3x+1}{(x-1)(x+3)}$

(c) $\frac{4x+17}{(x+5)(x+4)}$ (d) $\frac{4x-10}{(x-2)(x-4)}$

(e) $\frac{5x+4}{(2x+1)(x+1)}$ (f) $\frac{x+1}{x(1-2x)}$

(g) $\frac{3x+7}{(x+1)^2}$ (h) $\frac{x}{(x-1)^2}$

Chapter 10

Self-assessment questions 10.1

1. To factorise an expression means to write it as a product, usually of two or more simpler expressions.

Exercise 10.1

1. (a) $9x+27$ (b) $-5x+10$ (c) $\frac{1}{2}x+\frac{1}{2}$
   (d) $-a+3b$ (e) $1/(2x+2y)$
   (f) $x/(yx-y^2)$
2. (a) $4x^2$ has factors 1, 2, 4, x, $2x$, $4x$, x^2, $2x^2$, $4x^2$;
   (b) $6x^3$ has factors 1, 2, 3, 6, x, $2x$, $3x$, $6x$, x^2, $2x^2$, $3x^2$, $6x^2$, x^3, $2x^3$, $3x^3$, $6x^3$
3. (a) $3(x+6)$ (b) $3(y-3)$ (c) $-3(y+3)$
   (d) $-3(1+3y)$ (e) $5(4+t)$ (f) $5(4-t)$
   (g) $-5(t+4)$ (h) $3(x+4)$ (i) $17(t+2)$
   (j) $4(t-9)$
4. (a) $x(x^3+2)$ (b) $x(x^3-2)$
   (c) $x(3x^3-2)$ (d) $x(3x^3+2)$
   (e) $x^2(3x^2+2)$ (f) $x^3(3x+2)$
   (g) $z(17-z)$ (h) $x(3-y)$ (i) $y(3-x)$
   (j) $x(1+2y+3yz)$
5. (a) $10(x+2y)$ (b) $3(4a+b)$
   (c) $2x(2-3y)$ (d) $7(a+2)$ (e) $5(2m-3)$
   (f) $1/[5(a+7b)]$ (g) $1/[5a(a+7b)]$
6. (a) $3x(5x+1)$ (b) $x(4x-3)$
   (c) $4x(x-2)$ (d) $3(5-x^2)$
   (e) $5x^2(2x+1+3y)$ (f) $6ab(a-2b)$
   (g) $8b(2ac-ba+3c)$

Self-assessment questions 10.2

1. $2x^2+x+6$ cannot be factorised. On the other hand, $2x^2+x-6$ can be factorised as $(2x-3)(x+2)$.

Exercise 10.2

1. (a) $(x+2)(x+1)$ (b) $(x+7)(x+6)$
   (c) $(x+5)(x-3)$ (d) $(x+10)(x-1)$
   (e) $(x-8)(x-3)$ (f) $(x-10)(x+10)$
   (g) $(x+2)(x+2)$ or $(x+2)^2$
   (h) $(x+6)(x-6)$ (i) $(x+5)(x-5)$
   (j) $(x+1)(x+9)$ (k) $(x+9)(x-1)$
   (l) $(x+1)(x-9)$ (m) $(x-1)(x-9)$
   (n) $x(x-5)$
2. (a) $(2x+1)(x-3)$ (b) $(3x+1)(x-2)$
   (c) $(5x+3)(2x+1)$
   (d) $2x^2+12x+16$ has a common factor of 2, which should be written outside a bracket to give $2(x^2+6x+8)$. The bracket can then be factorised to give $2(x+4)(x+2)$
   (e) $(2x+3)(x+1)$ (f) $(3s+2)(s+1)$
   (g) $(3z+2)(z+5)$
   (h) $9(x^2-4)=9(x+2)(x-2)$
   (i) $(2x+5)(2x-5)$
3. (a) $(x+y)(x-y)=x^2+yx-yx-y^2$ $=x^2-y^2$
   (b) (i) $(4x+1)(4x-1)$ (ii) $(4x+3)(4x-3)$
   (iii) $(5t-4r)(5t+4r)$
4. (a) $(x-2)(x+5)$ (b) $(2x+5)(x-4)$

(c) $(3x+1)(3x-1)$ (d) $10x^2+14x-12$ has a common factor of 2, which is written outside a bracket to give $2(5x^2+7x-6)$. The bracket can then be factorised to give $2(5x-3)(x+2)$; (e) $(x+13)(x+2)$ (f) $(-x+1)(x+3)$

5. (a) $(10+7x)(10-7x)$
   (b) $(6x+5y)(6x-5y)$
   (c) $\left(\frac{1}{2}+3v\right)\left(\frac{1}{2}-3v\right)$ (d) $\left(\frac{x}{y}+2\right)\left(\frac{x}{y}-2\right)$

Chapter 11

Self-assessment questions 11.1

1. The subject of a formula is a variable that appears by itself on one side of the formula. It appears nowhere else.
2. To change the subject of a formula you will have to transpose it.
3. Whatever is done to the left-hand side (l.h.s.) must be done to the right-hand side (r.h.s.). We can (a) add or subtract the same quantity from both left- and right-hand sides, (b) multiply or divide both sides by the same quantity, (c) perform the same operation on both sides.

Exercise 11.1

1. (a) $x=\frac{y}{3}$ (b) $x=\frac{1}{y}$ (c) $x=\frac{5+y}{7}$
   (d) $x=2y+14$ (e) $x=\frac{1}{2y}$
   (f) $2x+1=\frac{1}{y}$ so $2x=\frac{1}{y}-1=\frac{1-y}{y}$
   finally, $x=\frac{1-y}{2y}$
   (g) $y-1=\frac{1}{2x}$ so $2x=\frac{1}{y-1}$
   and then $x=\frac{1}{2(y-1)}$
   (h) $x=\frac{y+21}{18}$ (i) $x=\frac{19-y}{8}$
2. (a) $m=\frac{y-c}{x}$ (b) $x=\frac{y-c}{m}$
   (c) $c=y-mx$
3. (a) $y=13x-26$ and so $x=\frac{y+26}{13}$
   (b) $y=x+1$ and so $x=y-1$
   (c) $y=a+tx-3t$ and so $tx=y+3t-a$ and then $x=\frac{y+3t-a}{t}$
4. (a) $x=\frac{y-11}{7}$ (b) $I=\frac{V}{R}$ (c) $r^3=\frac{3V}{4\pi}$ and so $r=\sqrt[3]{\frac{3V}{4\pi}}$ (d) $m=\frac{F}{a}$
5. $n=(l-a)/d+1$
6. $\sqrt{x}=(m-n)/t$ and so $x=[(m-n)/t]^2$
7. (a) $x^2=1-y$ and so $x=\pm\sqrt{1-y}$
   (b) $1-x^2=1/y$ and so $x^2=1-1/y$, finally, $x=\pm\sqrt{1-1/y}$
   (c) Write $(1+x^2)y=1-x^2$, remove brackets to get $y+x^2y=1-x^2$, rearrange and factorise to give $x^2(y+1)=1-y$ from which $x^2=(1-y)/(1+y)$, finally $x=\pm\sqrt{(1-y)/(1+y)}$

Chapter 12

Self-assessment questions 12.1

1. A root is a solution of an equation.
2. A linear equation can be written in the form $ax + b = 0$, where a and b are constants. In a linear equation, the variable is always to the first power only.
3. Adding or subtracting the same quantity from both sides; multiplying or dividing both sides by the same quantity; performing the same operation on both sides of the equation.
4. Equation: The two sides of the equation are equal only for certain values of the variable involved. These values are the solutions of the equation. For all other values, the two sides are not equal. Formula: A formula is essentially a statement of how two, or more, variables relate to one another, as for example in $A = \pi r^2$. The area of a circle, A, is always πr^2, for any value of the radius r.

Exercise 12.1

2. (a) $x = \frac{9}{3} = 3$ (b) $x = 3 \times 9 = 27$
   (c) $3t = -6$, so $t = \frac{-6}{3} = -2$ (d) $x = 22$
   (e) $3x = 4$ and so $x = \frac{4}{3}$ (f) $x = 36$
   (g) $5x = 3$ and so $x = \frac{3}{5}$ (h) $x + 3 = 6$ and so $x = 3$
   (i) adding the two terms on the left we obtain

$$\frac{3x+2}{2} + 3x = \frac{3x+2}{2} + \frac{6x}{2}$$

$$= \frac{3x+2+6x}{2} = \frac{9x+2}{2}$$

   and the equation becomes $(9x + 2)/2 = 1$, therefore $9x + 2 = 2$, that is, $9x = 0$ and so $x = 0$

3. (a) $5x + 10 = 13$, so $5x = 3$ and so $x = \frac{3}{5}$
   (b) $3x - 21 = 2x + 2$ and so $x = 23$
   (c) $5 - 10x = 8 - 4x$ so that $-3 = 6x$ finally $x = \frac{-3}{6} = -\frac{1}{2}$
4. (a) $t = 9$ (b) $v = \frac{17}{7}$
   (c) $3s + 2 = 14s - 14$, so that $11s = 16$ and so $s = \frac{16}{11}$
5. (a) $t = 3$ (b) $t = 5$ (c) $t = -5$ (d) $t = 3$
   (e) $x = \frac{12}{5}$ (f) $x = -9$ (g) $x = 24$
   (h) $x = 15$ (i) $x = 23$ (j) $x = \frac{43}{19}$
6. (a) $x = \frac{1}{5}$ (b) $x = \frac{2}{5}$ (c) $x = -\frac{3}{5}$
   (d) $x = 1\frac{2}{5}$ (e) $x = -1$ (f) $x = -3/5$
   (g) $x = -1/2$ (h) $x = -1/10$

Exercise 12.2

2. (a) Adding the two equations gives

$$\begin{aligned} 3x + y &= 1 \\ 2x - y &= 2 \\ \hline 5x \quad &= 3 \end{aligned}$$

   from which $x = \frac{3}{5}$. Substitution into either equation gives $y = -\frac{4}{5}$.
   (b) Subtracting the given equations eliminates y to give $x = 4$. From either equation $y = 1$.
   (c) Multiplying the second equation by 2 and subtracting the first gives

$$\begin{aligned} 2x + 6y &= 24 \\ 2x - y &= 17 \quad - \\ \hline 7y &= 7 \end{aligned}$$

   from which $y = 1$. Substitute into either equation to get $x = 9$.

(d) Multiplying the second equation by -2 and subtracting from the first gives

$$\begin{aligned} -2x + y &= -21 \\ -2x - 6y &= 28 \quad - \\ \hline 7y &= -49 \end{aligned}$$

from which $y = -7$. Substitution into either equation gives $x = 7$.

(e) Multiplying the first equation by 3 and adding the second gives

$$\begin{aligned} -3x + 3y &= -30 \\ 3x + 7y &= 20 \quad + \\ \hline 10y &= -10 \end{aligned}$$

from which $y = -1$. Substitution into either equation gives $x = 9$.

(f) Multiplying the second equation by 2 and subtracting this from the first gives

$$\begin{aligned} 4x - 2y &= 2 \\ 6x - 2y &= 8 \quad - \\ \hline -2x \quad &= -6 \end{aligned}$$

from which $x = 3$. Substitution into either equation gives $y = 5$.

3. (a) $x = 7, y = 1$ (b) $x = -7, y = 2$ (c) $x = 10, y = 0$ (d) $x = 0, y = 5$ (e) $x = -1, y = -1$

Self-assessment questions 12.2

1. If $b^2 - 4ac > 0$ a quadratic equation will have distinct real roots.
2. If $b^2 - 4ac = 0$ a quadratic equation will have a repeated root.

Exercise 12.3

1. (a) $(x + 2)(x - 1) = 0$ so that $x = -2$ and 1
(b) $(x - 3)(x - 5) = 0$ so that $x = 3$ and 5
(c) $2(2x + 1)(x + 1) = 0$ so that $x = -1$ and $x = -\frac{1}{2}$
(d) $(x - 3)(x - 3) = 0$ so that $x = 3$ twice
(e) $(x + 9)(x - 9) = 0$ so that $x = -9$ and 9
(f) $-1, -3$ (g) $1, -3$ (h) $1, -4$
(i) $-1, -5$ (j) $5, 7$ (k) $-5, -7$
(l) $1, -\frac{3}{2}$ (m) $-\frac{3}{2}, 2$ (n) $-\frac{3}{2}, 5$
(o) $1, -1/3$ (p) $-1/3, 5/3$ (q) $0, -\frac{1}{7}$
(r) $-\frac{3}{2}$ twice

2. (a) $$x = \frac{-(-6) \pm \sqrt{(-6)^2 - 4(3)(-5)}}{6} = \frac{6 \pm \sqrt{96}}{6} = 2.633 \text{ and } -0.633$$

(b) $$x = \frac{-3 \pm \sqrt{9 - 4(1)(-77)}}{2} = \frac{-3 \pm \sqrt{317}}{2} = 7.402 \text{ and } -10.402$$

(c) $$x = \frac{-(-9) \pm \sqrt{(-9)^2 - 4(2)(2)}}{4} = \frac{9 \pm \sqrt{65}}{4} = 4.266 \text{ and } 0.234$$

(d) $1, -4$ (e) $1.758, -0.758$
(f) $0.390, -0.640$ (g) $7.405, -0.405$
(h) $0.405, -7.405$ (i) no solutions
(j) $2.766, -1.266$

3. (a) $(3x + 2)(2x + 3) = 0$ from which $x = -2/3$ and $-3/2$
(b) $(t + 3)(3t + 4) = 0$ from which $t = -3$ and $t = -4/3$
(c) $t = (7 \pm \sqrt{37})/2 = 6.541$ and 0.459

Self-assessment questions 12.3

1. $y = mx + c$
2. m is the gradient of the line. c is the intercept of the line with the vertical axis.

Exercise 12.4

1. (a), (b), (c) and (e) are straight lines.
2. (a) $m = 9$, $c = -11$ (b) $m = 8$, $c = 1.4$
   (c) $m = \frac{1}{2}$, $c = -11$ (d) $m = -2$, $c = 17$
   (e) $m = \frac{2}{3}$, $c = \frac{1}{3}$ (f) $m = -\frac{2}{5}$, $c = \frac{4}{5}$
   (g) $m = 3$, $c = -3$ (h) $m = 0$, $c = 4$
3. (a) (i) -1 (ii) 6 (b) (i) 2 (ii) -1
   (c) (i) 2 (ii) -1.5 (d) (i) $\frac{3}{4}$ (ii) 3
   (e) (i) -6 (ii) 18

Self-assessment questions 12.4

1. gradient = m =

$$\frac{\text{difference in } y \text{ (vertical) coordinates}}{\text{difference in } x \text{ (horizontal) coordinates}}$$

$$m = \frac{y_2 - y_1}{x_2 - x_1}$$

2. The value of c is given by the intersection of the straight line graph with the vertical axis.

Exercise 12.5

1. $y = 2x + 5$

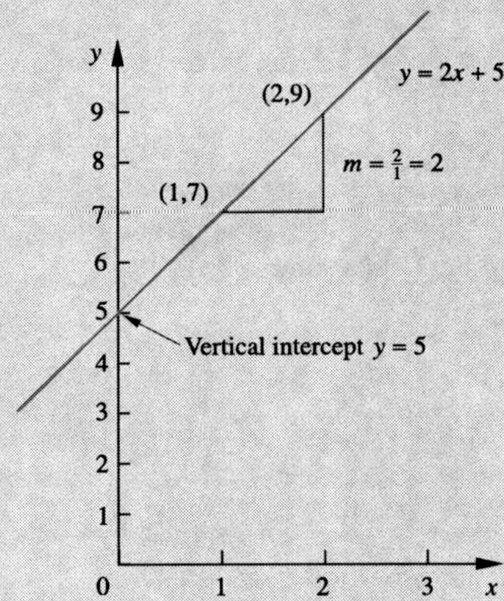

2. $y = 6x - 10$
3. $y = 2$
4. $y = x$
5. $y = -x$
6. $y = 2x + 8$
7. (b), (c) and (d) lie on the line.
8. The equation of the line is $y = mx + c$. The gradient is $m = -1$ and so

   $$y = -x + c$$

   When $x = -3$, $y = 7$ and so $c = 4$. Hence the required equation is $y = -x + 4$.
9. The gradient of $y = 3x + 17$ is 3. Let the required equation be $y = mx + c$.
   Since the lines are parallel then $m = 3$ and so $y = 3x + c$. When $x = -1$, $y = -6$ and so $c = -3$. Hence the equation is $y = 3x - 3$.
10. The equation is $y = mx + c$.
    The vertical intercept is $c = -2$ and so

    $$y = mx - 2$$

    When $x = 3$, $y = 10$ and so $m = 4$. Hence the required equation is $y = 4x - 2$.

Self-assessment questions 12.5

1. A tangent is a straight line that touches a curve at a single point.
2. The gradient of a curve at a particular point is the gradient of the tangent at that point.

Exercise 12.6

1.

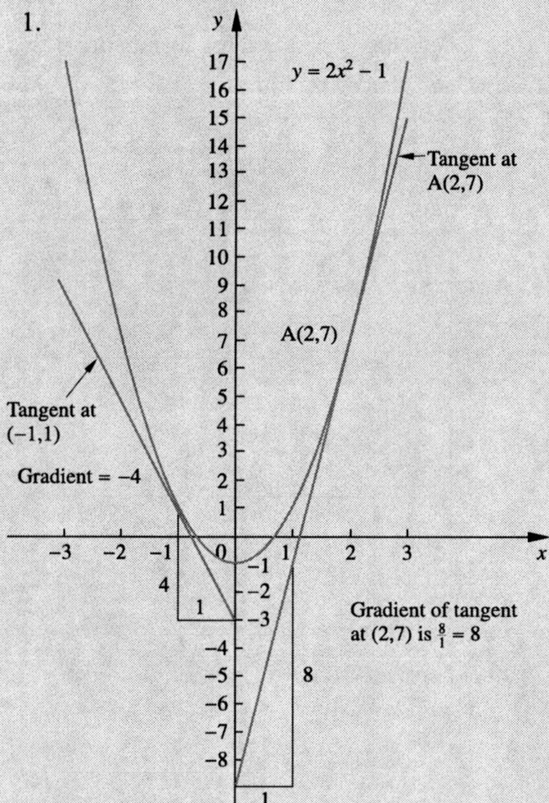

2.

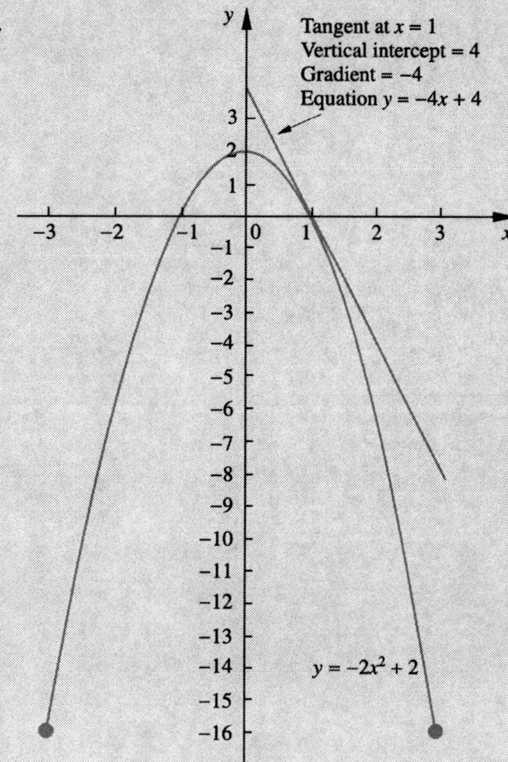

Exercise 12.7

1. (a) 9.9742; (b) 6.6859; (c) 0.5488; (d) 0.8808; (e) −8.4901.
2. (a) $e^2.e^7 = e^{2+7} = e^9$ (b) $\frac{e^7}{e^4} = e^{7-4} = e^3$
   (c) $\frac{e^{-2}}{e^3} = e^{-2-3} = e^{-5}$ or $\frac{1}{e^5}$
   (d) $\frac{e^3}{e^{-1}} = e^{3-(-1)} = e^4$;
   (e) $\frac{(4e)^2}{(2e)^3} = \frac{4^2e^2}{2^3e^3} = \frac{16e^{2-3}}{8} = 2e^{-1} = \frac{2}{e}$
   (f) $e^2(e + \frac{1}{e}) - e = e^2e + \frac{e^2}{e} - e$
   $= e^3 + e - e = e^3$
   (g) $e^{1.5}e^{2.7} = e^{1.5+2.7} = e^{4.2}$
   (h) $\sqrt{e}\sqrt{4e} = e^{0.5}(4e)^{0.5} = e^{0.5}4^{0.5}e^{0.5}$
   $= 2e^1 = 2e$

3. (a) $e^x e^{-3x} = e^{x-3x} = e^{-2x}$
   (b) $e^{3x} e^{-x} = e^{3x-x} = e^{2x}$
   (c) $e^{-3x} e^{-x} = e^{-3x-x} = e^{-4x}$
   (d) $(e^{-x})^2 e^{2x} = e^{-2x} e^{2x} = e^0 = 1$
   (e) $\frac{e^{3x}}{e^x} = e^{3x-x} = e^{2x}$
   (f) $\frac{e^{-3x}}{e^{-x}} = e^{-3x-(-x)} = e^{-2x}$
4. (a) $\frac{e^t}{e^3} = e^{t-3}$ (b) $\frac{2e^{3x}}{4e^x} = \frac{1}{2}e^{3x-x} = \frac{e^{2x}}{2}$
   (c) $\frac{(e^x)^3}{3e^x} = \frac{e^{3x}}{3e^x} = \frac{e^{3x-x}}{3} = \frac{e^{2x}}{3}$
   (d) $e^x + 2e^{-x} + e^x = 2e^x + 2e^{-x}$
   $= 2(e^x + e^{-x})$
   (e) $\frac{e^x e^y e^z}{e^{(x/2)-y}} = e^{x+y+z-((\frac{x}{2})-y)} = e^{(\frac{x}{2})+2y+z}$
   (f) $\frac{(e^{\frac{t}{3}})^6}{(e^t + e^t)} = \frac{e^{(\frac{t}{3})\times 6}}{2e^t} = \frac{e^{2t}}{2e^t} = \frac{e^t}{2}$
   (g) $\frac{e^t}{e^{-t}} = e^{t-(-t)} = e^{2t}$
   (h) $\frac{1+e^{-t}}{e^{-t}} = (1+e^{-t})e^t = e^t + e^{-t}e^t$
   $= e^t + e^0 = e^t + 1$
   (i) $(e^z e^{-\frac{z}{2}})^2 = (e^{z-\frac{z}{2}})^2 = (e^{\frac{z}{2}})^2 = e^z$
   (j) $\frac{e^{-3+t}}{2e^t} = \frac{e^{-3+t-t}}{2} = \frac{e^{-3}}{2} = \frac{1}{2e^3}$
   (k) $(\frac{e^{-1}}{e^{-x}})^{-1} = (e^{-1-(-x)})^{-1} = (e^{x-1})^{-1}$
   $= e^{(x-1)(-1)} = e^{1-x}$

Self-assessment questions 12.6

1. The functions are never negative. When $x = 0$ the functions have a value of 1.
2. False; e^{-x} is positive for all values of x.

Exercise 12.8

1.

x	-1	-0.8	-0.6	-0.4	-0.2
e^{3x}	0.0498	0.0907	0.1653	0.3012	0.5488

x	0	0.2	0.4	0.6	0.8	1
e^{3x}	1	1.8221	3.3201	6.0496	11.0232	20.0855

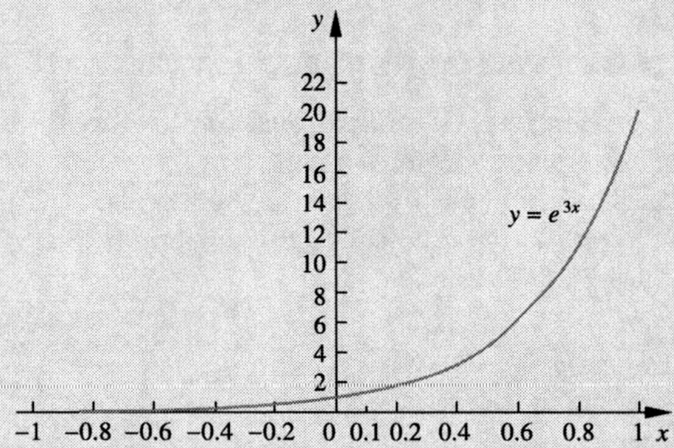

The graph has similar shape and properties to that of $y = e^x$.

2. (a)

t	0	1	2	3	4	5
P	5	8.1606	9.3233	9.7511	9.9084	9.9663

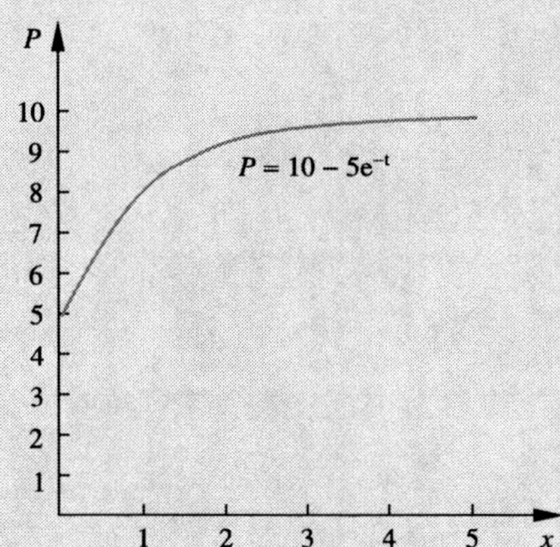

(b) Maximum population is 10, obtained for large values of t.

3. (a)

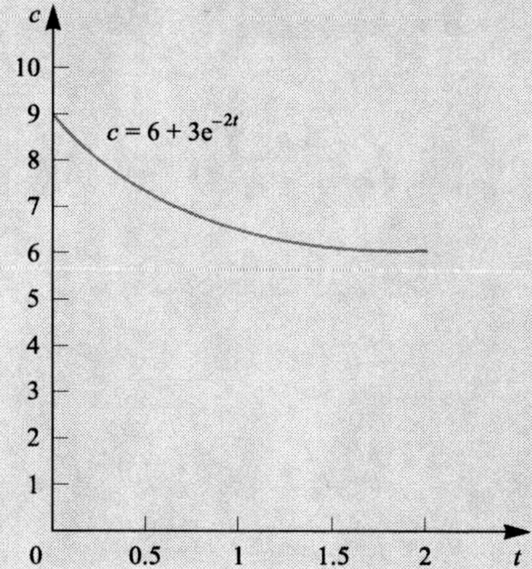

(b) 6

4. (a) $\cosh x + \sinh x$

$$= \frac{e^x + e^{-x}}{2} + \frac{e^x - e^{-x}}{2}$$

$$= \frac{e^x + e^{-x} + e^x - e^{-x}}{2} = e^x$$

(b) $\cosh x - \sinh x$

$$= \frac{e^x + e^{-x}}{2} - \frac{(e^x - e^{-x})}{2}$$

$$= \frac{e^x + e^{-x} - e^x + e^{-x}}{2} = e^{-x}$$

5. $3 \sinh x + 7 \cosh x$

$$= 3\left(\frac{e^x - e^{-x}}{2}\right) + 7\left(\frac{e^x + e^{-x}}{2}\right)$$

$$= 5e^x + 2e^{-x}$$

6. $6e^x - 9e^{-x} = 6(\cosh x + \sinh x)$

$$-9(\cosh x - \sinh x)$$

$$= 15 \sinh x - 3 \cosh x$$

Exercise 12.9

1. (a)

x	0	0.5	1	1.5	2	2.5	3
$15 - x^2$	15	14.75	14	12.75	11	8.75	6
e^x	1	1.65	2.72	4.48	7.39	12.18	20.09

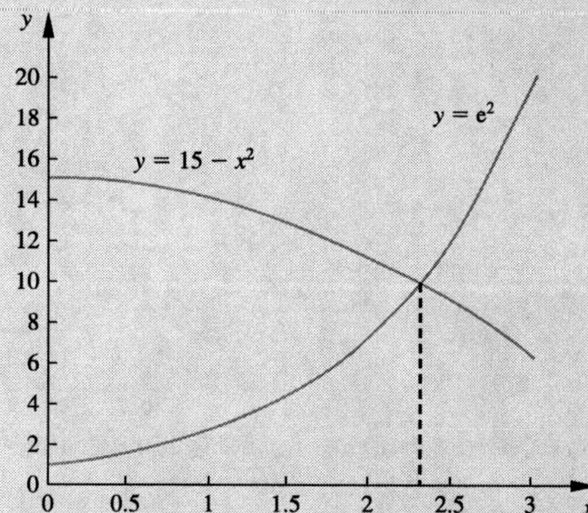

(b) Solving $e^x + x^2 = 15$ is equivalent to solving $e^x = 15 - x^2$. Approximate solution is $x = 2.3$.

2. (a)

x	−1	−0.75	−0.5	−0.25
$12x^2 - 1$	11	5.75	2	−0.25
$3e^{-x}$	8.15	6.35	4.95	3.85

x	0	0.25	0.5	0.75	1
$12x^2 - 1$	−1	−0.25	2	5.75	11
$3e^{-x}$	3	2.34	1.82	1.42	1.10

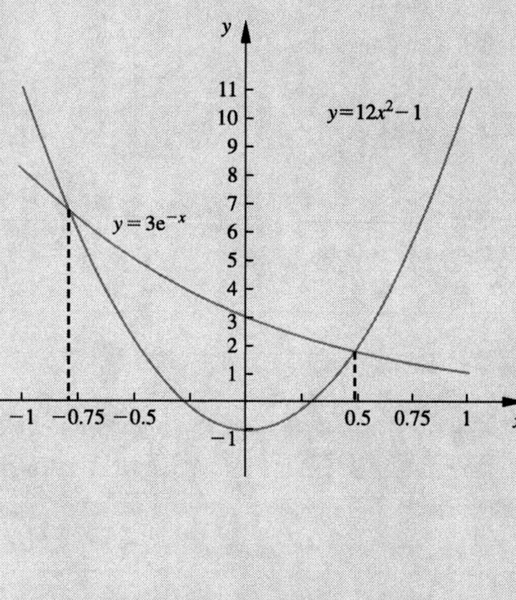

(b) Approximate solutions are $x = -0.80$ and $x = 0.49$.

3.

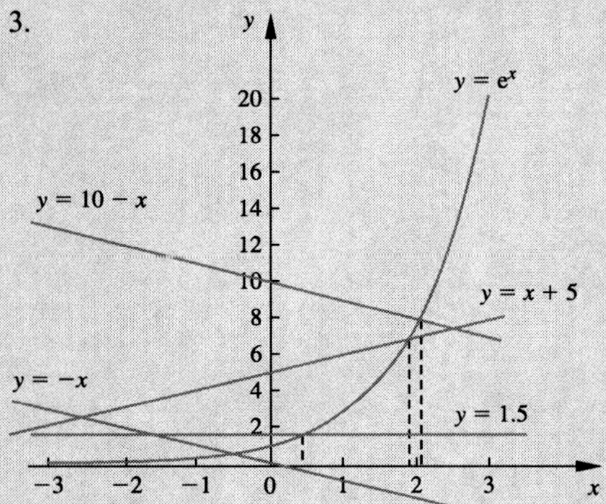

(a) The equation $e^x + x = 0$ is equivalent to $e^x = -x$. Hence draw $y = -x$. Intersection of $y = e^x$ and $y = -x$ will give solution to $e^x + x = 0$. Approximate solution: $x = -0.57$.

(b) The equation $e^x - 1.5 = 0$ is equivalent to $e^x = 1.5$. Hence also draw $y = 1.5$. This is a horizontal straight line. Intersection point gives solution to $e^x - 1.5 = 0$. Approximate solution: $x = 0.41$.

(c) The equation $e^x - x - 5 = 0$ can be written as $e^x = x + 5$. Hence draw $y = x + 5$. Approximate solution: $x = 1.94$.

(d) The equation

$$\frac{e^x}{2} + \frac{x}{2} - 5 = 0$$

can be written as

$$e^x = 10 - x$$

Hence also draw $y = 10 - x$. Approximate solution: $x = 2.07$.

4. (a)

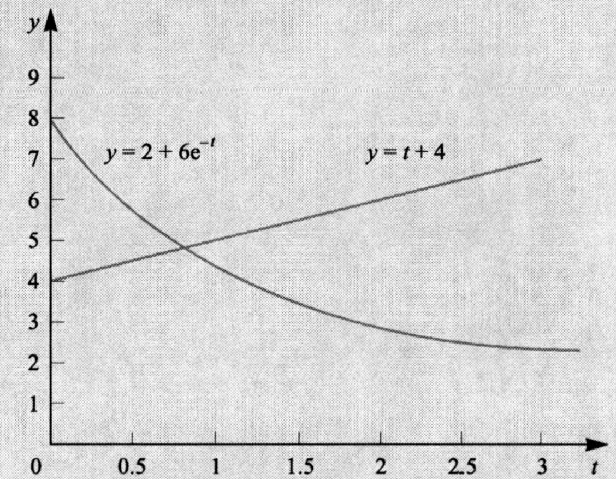

(b) The t value corresponding to $y = 5$ on the graph is $t = 0.7$. Hence the root of $2 + 6e^{-t} = 5$ is $t = 0.7$.

(c) $y = t + 4$ is also shown. The point of intersection of $y = 2 + 6e^{-t}$ and $y = t + 4$ is a solution of

$$2 + 6e^{-t} = t + 4$$

$$6e^{-t} = t + 2$$

The solution is $t = 0.77$.

Chapter 13

Self-assessment questions 13.1

1. False. For example, natural logarithms have base e (≈ 2.718).

Exercise 13.1

1. (a) 2.1761 (b) 5.0106 (c) −0.5003 (d) −2.3026

2. (a) $\log_3 6561 = 8$ (b) $\log_6 7776 = 5$
   (c) $\log_2 1024 = 10$
   (d) $\log_{10} 100000 = 5$ (e) $\log_4 16384 = 7$
   (f) $\log_{0.5} 0.03125 = 5$
   (g) $\log_{12} 1728 = 3$ (h) $\log_9 6561 = 4$
3. (a) $1296 = 6^4$ (b) $225 = 15^2$
   (c) $512 = 8^3$ (d) $2401 = 7^4$ (e) $243 - 3^5$
   (f) $216 = 6^3$ (g) $8000 = 20^3$
   (h) $4096 = 16^3$ (i) $4096 = 2^{12}$
4. (a) 0.4771
   (b) $$3 = 10^{0.4771}$$ $$3 \times 10^2 = 10^2 \times 10^{0.4771}$$ $$300 = 10^{2.4771}$$ $$\log 300 = 2.4771$$
   (c) $$3 = 10^{0.4771}$$ $$3 \times 10^{-2} = 10^{-2} \times 10^{0.4771}$$ $$0.03 = 10^{-1.5229}$$ $$\log 0.03 = -1.5229$$
5. (a) 7
   (b) $$\log 7 = 0.8451$$ $$7 = 10^{0.8451}$$ $$7 \times 10^2 = 10^2 \times 10^{0.8451}$$ $$700 = 10^{2.8451}$$ $$\log 700 = 2.8451$$
   (c) $$7 = 10^{0.8451}$$ $$10^{-2} \times 7 = 10^{-2} \times 10^{0.8451}$$ $$0.07 = 10^{-1.1549}$$ $$\log 0.07 = -1.1549$$

Exercise 13.2

1. (a) $\log_4 6 = \log 6/\log 4 = 1.2925$
   (b) $\log_3 10 = \log 10/\log 3 = 2.0959$
   (c) $\log_{20} 270 = \log 270/\log 20 = 1.8688$
   (d) $\log_5 0.65 = \log 0.65/\log 5 = -0.2677$
   (e) $\log_2 100 = \log 100/\log 2 = 6.6439$
   (f) $\log_2 0.03 = \log 0.03/\log 2 = -5.0589$
   (g) $\log_{100} 10 = \log 10/\log 100 = \frac{1}{2}$
   (h) $\log_7 7 = 1$
2. $$\ln X = \frac{\log_{10} X}{\log_{10} \mathrm{e}} = \frac{\log_{10} X}{0.4343} = 2.3026 \log_{10} X$$
3. (a) $\log_3 7 + \log_4 7 + \log_5 7$ $$= \frac{\log 7}{\log 3} + \frac{\log 7}{\log 4} + \frac{\log 7}{\log 5} = 4.3840$$
   (b) $$\log_8 4 + \log_8 0.25 = \frac{\log 4}{\log 8} + \frac{\log 0.25}{\log 8}$$ $$= 0.6667 + (-0.6667)$$ $$= 0$$
   (c) $$\log_{0.7} 2 = \frac{\log 2}{\log 0.7} = -1.9434$$
   (d) $$\log_2 0.7 = \frac{\log 0.7}{\log 2} = -0.5146$$

Self-assessment questions 13.2

1. $$\log A + \log B = \log AB$$ $$\log A - \log B = \log\left(\frac{A}{B}\right)$$ $$\log A^n = n \log A$$

Exercise 13.3

1. (a) $\log 5 + \log 9 = \log(5 \times 9) = \log 45$
   (b) $\log 9 - \log 5 = \log(\frac{9}{5}) = \log 1.8$
   (c) $\log 5 - \log 9 = \log(\frac{5}{9})$
   (d) $2 \log 5 + \log 1 = 2 \log 5 = \log 5^2$ $= \log 25$
   (e) $2 \log 4 - 3 \log 2 = \log 4^2 - \log 2^3$ $= \log 16 - \log 8 = \log(\frac{16}{8}) = \log 2$
   (f) $\log 64 - 2 \log 2 = \log(64/2^2) = \log 16$
   (g) $3 \log 4 + 2 \log 1 + \log 27 - 3 \log 12$ $= \log 4^3 + 2(0) + \log 27 - \log 12^3$

$$= \log\left(\frac{4^3 \times 27}{12^3}\right)$$

$= \log 1 = 0$

2. (a) $\log 3x$ (b) $\log 8x$ (c) $\log 1.5$
   (d) $\log T^2$ (e) $\log 10X^2$
3. (a) $3\ \log X - \log X^2 = 3\ \log X - 2\ \log X$
   $= \log X$
   (b) $\log y - 2\ \log \sqrt{y} = \log y - \log(\sqrt{y})^2$
   $= \log y - \log y = 0$
   (c) $5\ \log x^2 + 3\ \log \frac{1}{x} = 10\ \log x - 3\ \log x$
   $= 7\ \log x$
   (d) $4\ \log X - 3\ \log X^2 + \log X^3$
   $= 4\ \log X - 6\ \log X + 3\ \log X$
   $= \log X$
   (e) $3\ \log y^{1.4} + 2\ \log y^{0.4} - \log y^{1.2}$
   $= 4.2\ \log y + 0.8\ \log y - 1.2\ \log y$
   $= 3.8\ \log y$
4. (a) $\log 4x - \log x = \log(4x/x) = \log 4$
   (b) $\log t^3 + \log t^4 = \log(t^3 \times t^4) = \log t^7$
   (c) $\log 2t - \log\left(\frac{t}{4}\right) = \log\left(2t \div \frac{t}{4}\right)$
   $= \log\left(2t \times \frac{4}{t}\right) = \log 8$
   (d) $\log 2 + \log\left(\frac{3}{x}\right) - \log\left(\frac{x}{2}\right)$
   $= \log\left(2 \times \frac{3}{x}\right) - \log\left(\frac{x}{2}\right)$
   $= \log\left(\frac{6}{x} \div \frac{x}{2}\right)$
   $= \log\left(\frac{6}{x} \times \frac{2}{x}\right)$
   $= \log\left(\frac{12}{x^2}\right)$
   (e) $\log\left(\frac{t^2}{3}\right) + \log\left(\frac{6}{t}\right) - \log\left(\frac{1}{t}\right)$
   $= \log\left(\frac{t^2}{3} \times \frac{6}{t}\right) - \log\left(\frac{1}{t}\right)$
   $= \log(2t) - \log\left(\frac{1}{t}\right)$
   $= \log\left(\frac{2t}{1/t}\right)$
   $= \log 2t^2$
   (f) $2\ \log y - \log y^2 = \log y^2 - \log y^2 = 0$
   (g) $3\ \log\left(\frac{1}{t}\right) + \log t^2 = \log\left(\frac{1}{t}\right)^3 + \log t^2$
   $= \log\left(\frac{1}{t^3}\right) + \log t^2$
   $= \log\left(\frac{t^2}{t^3}\right)$
   $= \log\left(\frac{1}{t}\right) = -\log t$
   (h) $4\ \log \sqrt{x} + 2\ \log\left(\frac{1}{x}\right)$
   $= 4\ \log x^{0.5} + \log\left(\frac{1}{x}\right)^2$
   $= \log x^2 + \log\left(\frac{1}{x^2}\right)$
   $= \log\left(x^2\ \frac{1}{x^2}\right)$
   $= \log 1 = 0$
   (i) $2\ \log x + 3\ \log t = \log x^2 + \log t^3$
   $= \log(x^2 t^3)$

(j) $\log A - \frac{1}{2}\log 4A = \log A - \log(4A)^{1/2}$

$$= \log\left[\frac{A}{(4A)^{1/2}}\right]$$

$$= \log\left(\frac{A^{1/2}}{2}\right)$$

(k) $\dfrac{\log 9x + \log 3x^2}{3} = \dfrac{\log(9x.3x^2)}{3}$

$$= \frac{\log(27x^3)}{3}$$

$$= \frac{\log(3x)^3}{3}$$

$$= \frac{3\log(3x)}{3}$$

$$= \log(3x)$$

(l) $\log xy + 2\log\left(\frac{x}{y}\right) + 3\log\left(\frac{y}{x}\right)$

$$= \log xy + \log\left(\frac{x^2}{y^2}\right) + \log\left(\frac{y^3}{x^3}\right)$$

$$= \log\left(xy\frac{x^2}{y^2}\frac{y^3}{x^3}\right)$$

$$= \log\left(\frac{x^3y^4}{x^3y^2}\right)$$

$$= \log y^2$$

(m) $\log\left(\frac{A}{B}\right) - \log\left(\frac{B}{A}\right) = \log\left(\frac{A}{B} \div \frac{B}{A}\right)$

$$= \log\left(\frac{A}{B} \times \frac{A}{B}\right) = \log\left(\frac{A^2}{B^2}\right)$$

(n) $\log\left(\frac{2t}{3}\right) + \frac{1}{2}\log 9t - \log\left(\frac{1}{t}\right)$

$$= \log\left(\frac{2t}{3}\right) + \log(9t)^{1/2} - \log t^{-1}$$

$$= \log\left[\frac{2t}{3}(9t)^{1/2}\right] + \log t$$

$$= \log\left[\frac{2t}{3}(9t)^{1/2}t\right]$$

$$= \log\left(\frac{2t}{3}3t^{1/2}t\right)$$

$$= \log(2t^{2.5})$$

5. $\log_{10} X + \ln X = \log_{10} X + \dfrac{\log_{10} X}{\log_{10} \mathrm{e}}$

$$= \log_{10} X + 2.3026 \log_{10} X$$

$$= 3.3026 \log_{10} X$$

$$= \log X^{3.3026}$$

6. (a) $\log(9x-3) - \log(3x-1)$

$$= \log\left(\frac{9x-3}{3x-1}\right) = \log 3$$

(b) $\log(x^2-1) - \log(x+1)$

$$= \log\left(\frac{x^2-1}{x+1}\right) = \log(x-1)$$

(c) $\log(x^2+3x) - \log(x+3)$

$$= \log\left(\frac{x^2+3x}{x+3}\right) = \log x$$

Exercise 13.4

1. (a) $\log x = 1.6000$

$$x = 10^{1.6000} = 39.8107$$

(b) $10^x = 75$

$$x = \log 75 = 1.8751$$

(c) $\ln x = 1.2350$

$$x = e^{1.2350} = 3.4384$$

(d) $e^x = 36$

$$x = \ln 36 = 3.5835$$

2. (a) $\log(3t) = 1.8$

$$3t = 10^{1.8}$$

$$t = \frac{10^{1.8}}{3} = 21.0319$$

(b) $10^{2t} = 150$

$$2t = \log 150$$

$$t = \frac{1}{2} \log 150 = 1.0880$$

(c) $\ln(4t) = 2.8$

$$4t = e^{2.8}$$

$$t = \frac{1}{4} e^{2.8} = 4.1112$$

(d) $e^{3t} = 90$

$$3t = \ln 90$$

$$t = \frac{1}{3} \ln 90 = 1.4999$$

3. (a) $\log x = 0.3940$

$$x = 10^{0.3940} = 2.4774$$

(b) $\ln x = 0.3940$

$$x = e^{0.3940} = 1.4829$$

(c) $10^y = 5.5$

$$y = \log 5.5 = 0.7404$$

(d) $e^z = 500$

$$z = \ln 500 = 6.2146$$

(e) $\log(3v) = 1.6512$

$$3v = 10^{1.6512}$$

$$v = \frac{10^{1.6512}}{3} = 14.9307$$

(f) $\ln\left(\frac{t}{6}\right) = 1$

$$\frac{t}{6} = e^1 = e$$

$$t = 6e = 16.3097$$

(g) $10^{2r+1} = 25$

$$2r + 1 = \log 25$$

$$2r = \log 25 - 1$$

$$r = \frac{\log 25 - 1}{2} = 0.1990$$

(h) $e^{(2t-1)/3} = 7.6700$

$$\frac{2t - 1}{3} = \ln 7.6700$$

$$2t - 1 = 3 \ln 7.6700$$

$$2t = 3 \ln 7.6700 + 1$$

$$t = \frac{3 \ln 7.6700 + 1}{2} = 3.5560$$

(i) $\log(4b^2) = 2.6987$

$$4b^2 = 10^{2.6987}$$

$$b^2 = \frac{10^{2.6987}}{4} = 124.9223$$

$$b = \sqrt{124.9223} = \pm 11.1769$$

(j) $\log\left(\frac{6}{2+t}\right) = 1.5$

$$\frac{6}{2+t} = 10^{1.5}$$

$$6 = 10^{1.5}(2 + t)$$

$$\frac{6}{10^{1.5}} = 2 + t$$

$$t = \frac{6}{10^{1.5}} - 2 = -1.8103$$

(k) $\ln(2r^3 + 1) = 3.0572$

$$2r^3 + 1 = e^{3.0572}$$

$$2r^3 = e^{3.0572} - 1$$

$$r^3 = \frac{e^{3.0572} - 1}{2} = 10.1340$$

$$r = 2.1640$$

(l) $\ln(\log t) = -0.3$

$$\log t = e^{-0.3} = 0.7408$$

$$t = 10^{0.7408} = 5.5058$$

(m) $10^{t^2 - 1} = 180$

$$t^2 - 1 = \log 180$$

$$t^2 = \log 180 + 1 = 3.2553$$

$$t = \sqrt{3.2553} = \pm 1.8042$$

(n) $10^{3r^4} = 170000$

$$3r^4 = \log 170000$$

$$r^4 = \frac{\log 170000}{3} = 1.7435$$

$$r = \pm 1.1491$$

(o) $\log 10^t = 1.6$

$$t \log 10 = 1.6$$

$$t = 1.6$$

(p) $\ln(e^x) = 20000$

$$x \ln e = 20000$$

$$x = 20000$$

4. (a) $e^{3x}e^{2x} = 59$

$$e^{5x} = 59$$

$$5x = \ln 59$$

$$x = \frac{\ln 59}{5} = 0.8155$$

(b) $10^{3t}.10^{4-t} = 27$

$$10^{4+2t} = 27$$

$$4 + 2t = \log 27$$

$$2t = \log 27 - 4$$

$$t = \frac{\log 27 - 4}{2} = -1.2843$$

(c) $\log(5 - t) + \log(5 + t) = 1.2$

$$\log(5 - t)(5 + t) = 1.2$$

$$\log(25 - t^2) = 1.2$$

$$25 - t^2 = 10^{1.2}$$

$$t^2 = 25 - 10^{1.2}$$

$$= 9.1511$$

$$t = \pm 3.0251$$

(d) $\log x + \ln x = 4$

$$\log x + \frac{\log x}{\log e} = 4$$

$$\log x \left(1 + \frac{1}{\log e}\right) = 4$$

$$3.3026 \log x = 4$$

$$\log x = \frac{4}{3.3026} = 1.2112$$

$$x = 10^{1.2112} = 16.2619$$

Self-assessment questions 13.3

1. Both $\log x$ and $\ln x$ tend to infinity as x tends to infinity. Both have a value of 0 when $x = 1$.
2. True. The graph of the logarithm function increases without bound, albeit very slowly.

Exercise 13.5

1. (a)

x	0.5	1	2	3	4
$\log x^2$	−0.6021	0	0.6021	0.9542	1.2041

x	5	6	7	8	9	10
$\log x^2$	1.3979	1.5563	1.6902	1.8062	1.9085	2

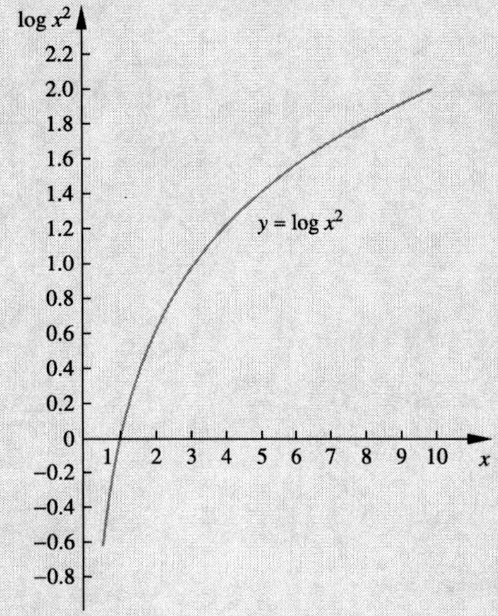

(b)

x	0.1	0.2	0.5	0.75	1
$\log(1/x)$	1	0.6990	0.3010	0.1249	0

x	2	3	4	5
$\log(1/x)$	−0.3010	−0.4771	−0.6021	−0.6990

x	6	7	8	9	10
$\log(1/x)$	−0.7782	−0.8451	−0.9031	−0.9542	−1

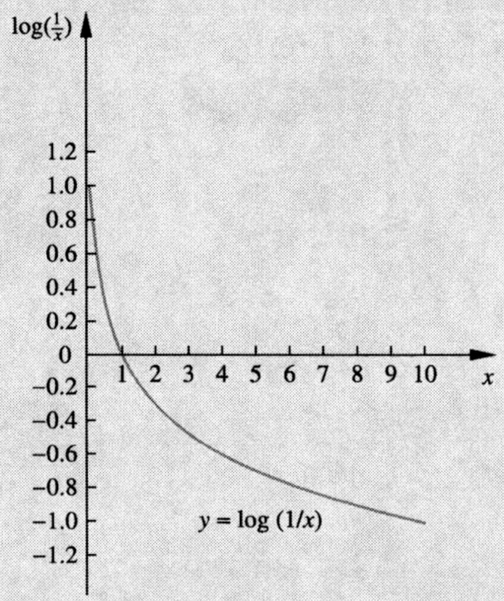

2. (a)

x	0.1	0.5	1	1.5	2	2.5	3
$\log x$	−1	−0.3010	0	0.1761	0.3010	0.3979	0.4771

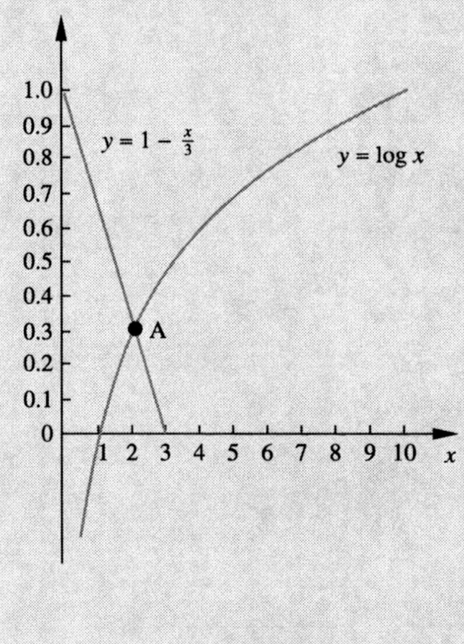

(b) $y = 1 - x/3$ is a straight line passing through (0,1) and (3,0).

(c) The graphs intersect at A. Approximate solution: $x = 2.06$.

3. (a)

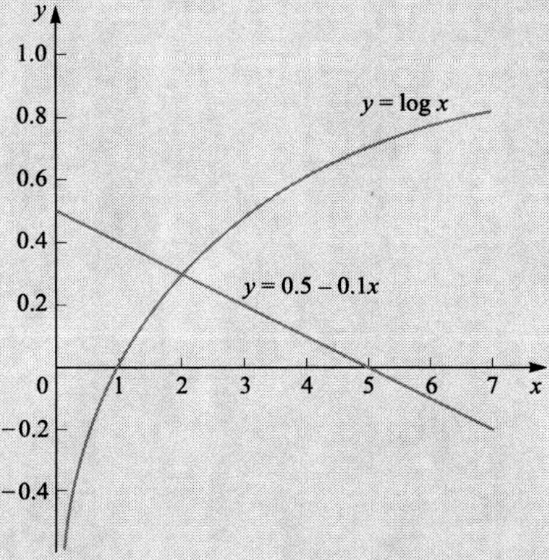

(b) The line $y = 0.5 - 0.1x$ is also shown on the figure.

(c) Solution $x = 2$.

Foundation Maths 4th Edition

Anthony Croft & Robert Davison

Solutions Manual

Solutions to Test & Assignment Exercises

Chapter 1. Arithmetic of whole numbers

1. a) 4 b) 2 c) 13 d) 4 e) 7 f) 9 g) 6 h) 4 i) 2 j) 3 k) 7 l) 6 m) 6 n) 3 o) 2
2. a) $2 \times 2 \times 2 \times 7$ b) 3×13 c) 2×37
3. a) 2 b) 6 c) 2 d) 8
4. a) 30 b) 143 c) 112 d) 120

Chapter 2. Measurement.

1. a) 0.125 g b) 15000 g c) 0.0005 g
2. a) 6.5 cm. b) 0.0245 cm. c) 500 cm.
3. 7.5 km., 0.1255 km.
4. 8000 ml
5. 0.056 l
6. 480 litres
7. 327.25 cm^2
8. 213.14°
9. 31415.9 cm^3
10. Volume = 2144.7 cm^3, Area = 804.2 cm^2
11. Area = $(\frac{y_1+y_2}{2})h$, 8.5 cm^2
12. 7500 kg.

Chapter 4. Fractions

1. a) $\frac{11}{12}$ b) $\frac{11}{10}$ c) $\frac{1}{21}$ d) $\frac{43}{30}$ e) $8\frac{5}{12}$ f) $\frac{23}{24}$ g) $\frac{5}{12}$ h) $\frac{23}{60}$
2. a) $\frac{3}{8}$ b) $\frac{4}{9}$ c) $\frac{20}{77}$ d) $\frac{2}{7}$ e) $\frac{13}{8}$ f) $\frac{2}{3}$ g) $-\frac{16}{17}$ h) $\frac{3}{5}$
3. a) $\frac{7}{6}$ b) $\frac{3}{4}$ c) 2 d) $\frac{199}{240}$ e) $\frac{4}{5}$ f) $\frac{5}{4}$ g) $\frac{4}{3}$ h) $\frac{19}{30}$
4. a) $\frac{1}{4}$ b) $\frac{3}{40}$ c) $\frac{2}{5}$ d) $\frac{2}{13}$ e) $\frac{3}{11}$

Chapter 4. Decimal fractions

1. a) $\frac{37}{50}$ b) $\frac{24}{25}$ c) $\frac{1}{20}$ d) $\frac{1}{4}$
2. a) $2\frac{1}{2}$ b) $3\frac{1}{4}$ c) $3\frac{1}{8}$ d) $6\frac{7}{8}$
3. a) 0.317 b) 0.095 c) 0.904
4. a) 0.1 b) 0.098 c) 0.1
5. a) 9.51 b) 9.5 c) 10
6. a) 20.0 b) 20 c) 20

Chapter 5. Percentage and ratio

1. a) 0.08 b) 0.18 c) 0.65
2. a) 37.5% b) 79% c) 47.46% (2 d.p.)
3. 408.408
4. 156.25
5. 0.2125
6. $\frac{1}{15}$, $\frac{2}{15}$, $\frac{3}{10}$
7. £ 41.04
8. £ 4600 and £ 8050
9. 8.974, 23.333, 37.692
10. 20% and 30%

Chapter 6. Algebra

1. $44^3 = 85184$, $0.44^2 = 0.1936$, $32.5^3 = 34328.125$.
2. a) $xxxyyyyy = x^3y^5$. b) $\frac{xxx}{yyyy} = \frac{x^3}{y^4}$, c) $a^2baab = a^4b^2$.
3. If $x = 2$, $y = 5$, $z = 3$, then $4x^3yz^2 = 1440$.
4. 3.142cm.
5. a) $21^2 - 16^2 = 185$, b) $(21 - 16)^2 = 25$. In general, $a^2 - b^2 \neq (a - b)^2$.
6. When $x = 4$ and $y = -3$, then

a) $xy = -12$, b) $\dfrac{x}{y} = -\dfrac{4}{3} = -1.3333$, c) $\dfrac{x^2}{y^2} = \dfrac{16}{9} = 1.778$, d) $\left(\dfrac{x}{y}\right)^2 = \dfrac{16}{9} = 1.778$.

7. When $x = 7$, $2x(x + 4) = 14(11) = 154$.
8. When $x = 9$, $4x^2 + 7x = 387$.
9. When $x = -2$, $3x^2 - 7x + 12 = 38$.
10. When $x = -3$, $-x^2 - 11x + 1 = 25$.
11. $I = \dfrac{V}{R} = \dfrac{10}{0.01} = 1000$.
12. a) $A = 1$, b) $A = 1/2$, c) $A = 1/3$.
13. a) $y = \dfrac{1}{2}$, b) y is not defined, c) $y = 1/12$.
14. a) When n is even, $(-1)^n = 1$. b) When n is odd $(-1)^n = -1$.
15. a) When n is even, $(-1)^{n+1} = -1$. b) When n is odd, $(-1)^{n+1} = 1$.

Chapter 7. Indices

1. a) $\frac{z^5}{z^{-5}} = z^{10}$, b) $z^0 = 1$, c) $\frac{z^8 z^6}{z^{14}} = 1$.
2. a) $0.25^{1/2} = \pm 0.5$ b) $4096^{1/3} = 16$ c) $2601^{1/2} = \pm 51$ d) $16^{-1/2} = \pm 0.25$.
3. x^8.
4. 1
5. a) $(abc^2)^2 = a^2b^2c^4$ b) $(xy^2z^3)^2 = x^2y^4z^6$ c) $(8x^2)^{-3} = 8^{-3}x^{-6} = \frac{1}{512x^6}$
6. a) $5792 = 5.792 \times 10^3$ b) $98.4 = 9.84 \times 10^1$ c) $0.001 = 1 \times 10^{-3}$ d) $-66.667 = -6.6667 \times 10^1$.

Chapter 8. Simplifying algebraic expressions

1. a) $11x^2 + x$ b) $8 - 17y$ c) $a^2 + a^3 - 2b^2$.
2. a) $-3a^5b^3c$ b) $-x$.
3. a) $7a^2 + 19ab - 6b^2$ b) $x^3 + 2x^2y$ c) $x^3 - xy^2$.
4. a) $21x^2 - x - 2$ b) $-x^2 - 2x + 3$ c) $5x + x^2$ d) $56x^2 + 12x - 8$.
5. a) $6x - 4x^2$ b) $-2a^2 - 5ab - 3b^2$ c) $17x + 2$ d) $-5a + 31$ e) $2b + 2a$.

Chapter 9. Algebraic fractions

1. a) $\frac{5}{4+3b}$ b) $\frac{a}{6a+3b}$ c) $\frac{ab}{6a+3b}$
   d) $\frac{5}{8}$ e) $\frac{1}{13+y}$ f) $13 + y$.

2. a) $5ab$ b) $\frac{16}{3x}$ c) $\frac{3t-4}{2}$
   d) $\frac{8x+41}{3}$ e) $\frac{61}{6x}$ f) $\frac{xy+3x}{y}$
   g) $\frac{x^2y^2+1}{xy}$ h) $\frac{3x+y}{(x+y)(x-y)}$.

3. a) $\frac{25y}{63}$ b) $\frac{32}{15x}$ c) $\frac{9x^2+5y^2}{6xy}$ d) $\frac{3m+n}{2}$ e) $\frac{m-n}{2}$ f) $\frac{m+n}{2}$ g) $\frac{5s-9}{30}$.
4. x.

Chapter 10. Factorisation

1. a) $7(x+7)$ b) $11(11x+2y)$ c) $a(a+b)$ d) $b(a+b)$ e) $ab(b+a)$.
2. a) $(3x-2)(x+1)$ b) $(x+12)(x-12)$ c) $(s-3)(s-2)$ d) $(2y+5)(y-3)$.
3. a) $(1+x)(1-x)$ b) $(x+1)(x-1)$ c) $(3+x)(3-x)$ d) $(x+9)(x-9)$ e) $(5+y)(5-y)$.
4. a) $x^2+6x=x(x+6)$ b) $s^2+3s+2=(s+1)(s+2)$ c) $s^2+s-2=(s+2)(s-1)$
   d) $x^2+11x+28=(x+7)(x+4)$ e) $2x^2-17x-9=(2x+1)(x-9)$.

Chapter 11. Transposing formulae

1. a) $x=\dfrac{y+18}{13}$ b) $x=\dfrac{13}{y}-18$ c) $x=\dfrac{13}{y-18}$.
2. a) $x=14y-7$ b) $v=\pm\sqrt{\dfrac{2E}{m}}$ c) $y=\dfrac{8x-12}{13}$ d) $x=\pm\sqrt{2gy}$.
3. $x=\left(\dfrac{k}{v}\right)^2$.
4. $h=\dfrac{V}{\pi r^2}$.
5. $r=\dfrac{y-H}{C}$.
6. $a=\dfrac{2(s-ut)}{t^2}$.
7. $u=\pm\sqrt{v^2-2as}$.
8. $v=\sqrt[3]{\dfrac{k}{p}}$.

Chapter 12. Solving equations

1. a) $t=\dfrac{12}{11}$ b) $t=1$ c) $t=0$.
2. a) $x=4$ b) $x=108$ c) $x=4$ d) $x=11$ e) $x=7$ f) $x=-\frac{1}{2}$ g) $x=0$.
3. a) $y=4$ b) $u=-\dfrac{1}{3}$ c) $I=\dfrac{5}{4}$ d) $i=-\dfrac{16}{11}$ e) $x=-8$ f) $x=-\dfrac{1}{10}$.
4. a) $x=\pm3$ b) $x=-0.86$, $x=-8.14$ c) $x=\pm4$ d) $x=-8$, $x=9$
   e) $x=\dfrac{11}{2}$, $x=-4$ f) $x=-3$, $x=7$.
5. True.
6. a) $x=-5$, $y=3$ b) $x=5$, $y=8$ c) $x=9$, $y=10$.

Chapter 12. The straight line

1. a), b), d) and e) have straight line graphs.

   a) gradient=2, intercept= -11

   b) gradient=5, intercept=10

   d) gradient=3, intercept=-3

   e) gradient=1, intercept=3/2

2. $y = 7x + 4$.
3. $y = -2x + 3$.
4. $y = 5$. The line does not pass through $(2, 6)$.
5. The graph is shown below.

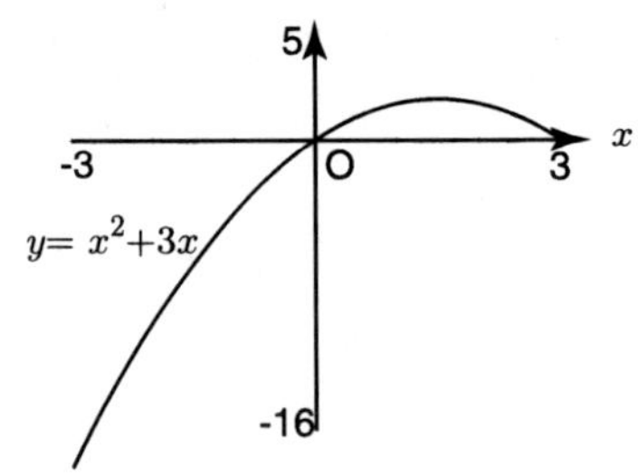

Graph for Q5

Gradient of tangent at $x = -2$ is 7. Gradient of tangent at $x = 1$ is 1.

6. a) $y = -2x$, b) $y = -4x$, c) $y = 4x$.
7. $y = 6x - 14$
8. $y = -6x + 3$
9. a) $x = -16/3$, b) $y = 32/5$
10. $y = \frac{3}{2}x + 3$
11. a) $y = -1$, b) $x = \frac{1}{4}$, c) at the point $(\frac{3}{4}, 2)$

Chapter 12. The exponential function

1. a) $e^{-3.1} = 0.045$ b) $e^{0.2} = 1.221$ c) $\frac{1}{e} = 0.368$.

2. The graph is shown below. Solution of $2e^{-x} - x = 0$ is approximately $x = 0.85$.

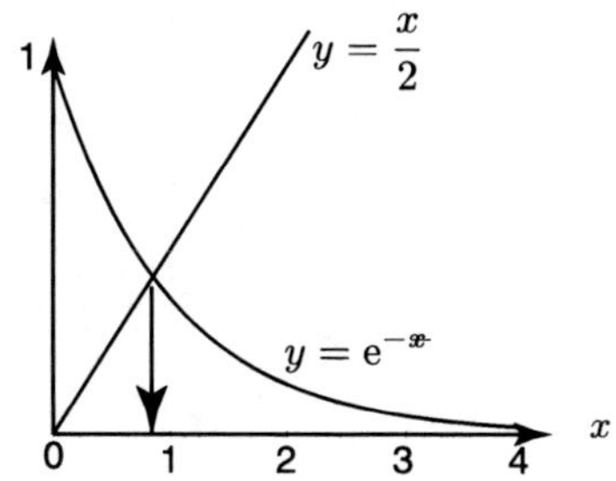

Graph for Q2

3. a) 1, b) e^x, c) e^{3x}, d) no simplification, e) $e^x + e^{3x}$

4. a) $e^{2x} - 2e^x + 1$, b) $e^{-2x} - 2e^{-x} + 1$, c) $e^{3x} + e^{2x}$, d) $2 + 6e^{2x}$

5. a) $e^{3t/2}$ b) e^6 c) $-e^{5x}$ d) e^{8-t} e) e^6 f) $\pm e^{2z}$ g) e^{3x} h) $\pm e^t(1+e)$, i) e^{-t}.

6. The graphs are shown below.

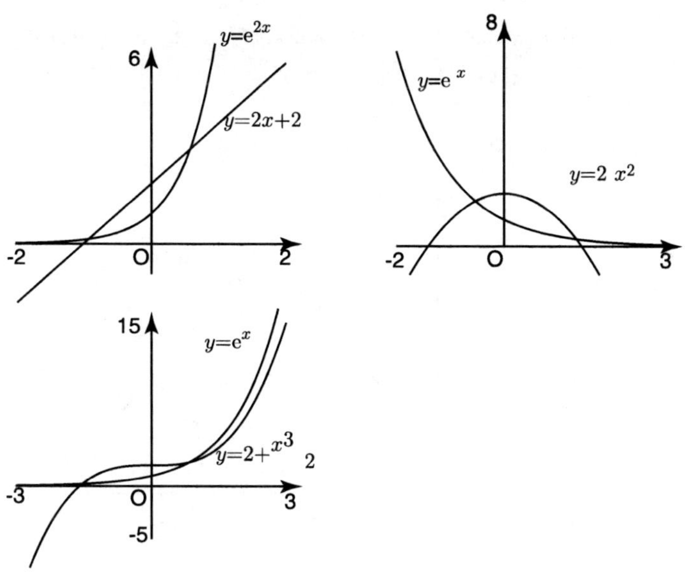

Graph for Q6 a) b) & c)

a) approximate solutions $x = -0.9$, $x = 0.6$.

b) approximate solutions $x = -0.5$, $x = 1.3$.

c) approximate solutions $x = -1.5$, $x = 0.8$.

7. The graph is shown below. Approximate solution $x = 0.9$.

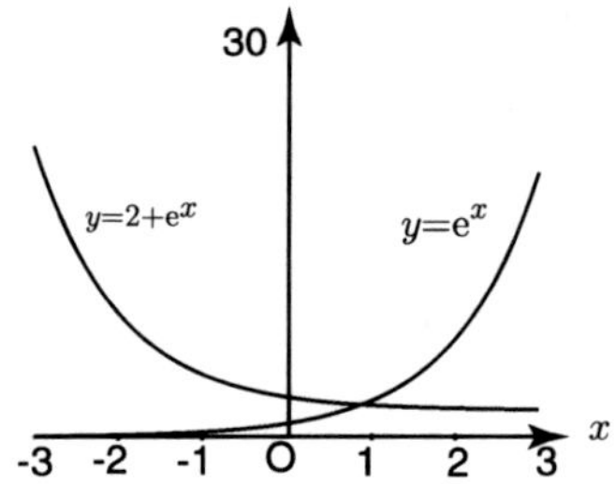

Graph for Q7

8. a) e^{8t} b) $e^{2t} - 1$ c) $e^{5y} + 1 + e^{3y}$ d) $1 + e^{2t} + 2e^t$.

9. $H \to 10$

10. $X \to 2$.

Chapter 13. The logarithm function

1. a) $\log 107000 = 5.029$ b) $\ln 0.0371 = -3.294$ c) $\log 0.1 = -1$ d) $\ln 150 = 5.011$.

2. a) 1.540 b) 1.661 c) 5.907 d) 2.410 e) 0.7737.

3. a) $\log 7t$ b) $\log \frac{t}{x}$ c) $\log xy^2$ d) $\log t^3 r^4$ e) $\log 3v^2$ f) $\log(a^2 - b^2)$ g) $\log xy$.

4. a) $x = 113.4755$ b) $x = 2.2408$ c) $x = 4.2777$ d) $x = 4.0275$, e) $t = 2.7358$ f) $t = 2.4028$ g) $x = 2.4105$

5. a), b) The graphs are shown below. c) Approximate solution: $x = 2.7$.

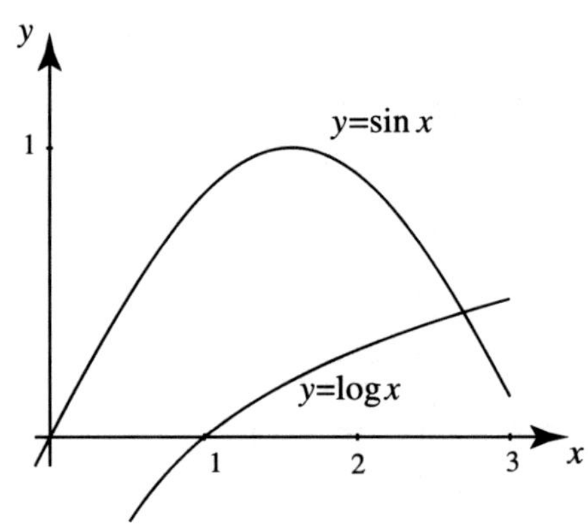

Graph for Q5

6. a) $x = 0.8451$, b) $x = 1.6073$, c) $x = 1.1990$, d) $x = -3.5215$, e) $x = 1.0095$

7. 0.6931, 1.9459

8. 0.6021, 0.6990

9. a) $x = \pm\sqrt{101}$, b) $x = 1.9542$, c) $x = 2$

10. a) $x = 0.3010, 0.6021$, b) $x = 0.3010$

Appendix 2
Answers

from *Get Ready for Biology*

Adapted from *Get Ready for A&P*

Lori K. Garrett

Answer Key

CHAPTER 1

Part A

1. –32
2. 20,000
3. 3/4
4. 1/4
5. 1/4
6. 0.2
7. 720 breaths per hour
8. 400 cm
9. 30.48 cm
10. 120 lbs

Part B

1. 14
2. 4
3. 0.3; 30%
4. "Normal" blood pressure is the average blood pressure.

Appendix 3
Answers to Questions That Test Your Understanding

Answers to questions that test your understanding:

Chapter 2 Measurement

1a) 1.2 mm
1b) 4 mm
1c) 3.5 mm
2a) 45 g
2b) 0.036 g
2c) 0.064 g
3a) 0.0035m^2
3b) 4.8 x 10^{-7} m^2
3c) 500 m^2
4a) 4000mm^3
4b) 8.53 x 10^{-4} mm^3
4c) 4 mm^3
5a) 34 ml
5b) 5000 ml
5c) 0.68 ml
6a) 108 acorns
6b) 290 km
6c) 12.5 g l^{-1}
6d) 6,667 trees
6e) 1.2 x 10^{10} cod
6f) 4 x 10^8 woodlice
7a) 905 μm^3
7b) 488.8 mm^3
7c) 551mm^2
8) 17.64 mm^2
9) 0.3136 m^2
10) 79.51 cm^3
11) 153.96 mm^2
12) 6.61 m^2
13) 0.204 μm^3
14) 796 m^3
15) 0.00021 m^3
16) 65.46 km^3
17) 1540 μm^3
18) 7.189 m^3
19) 36.1 cm^3
20) Top = 82.3 cm^3
Stalk = 6.44 cm^3
Biomass = 2.66 g
21) Volume = 17159.5 μm^3
Mass = 51478.5 ng
or 51.5 μg
22) Volume of bacterium = 0.0719 μm^3
Volume of colony = 3.595 x 10^7 μm^3
or 0.036 mm^3
23) Volume of worm = 4450 mm^3
Mass of worm = 1.78 g
24) Volume of one worm = 356.4 mm^3
Total volume of worms = 1.14 m^3

Chapter 3 Moles and Molarity

1) 0.048 moles
2) 56 g mol^{-1}
3) 0.168 mol l^{-1}
4) 0.377 mol l^{-1}
5) 0.1 l or 100 ml
6) NaCl = 4 ml
KOH = 4.5 ml
7) 0.2 mol
8) 0.25 mol l^{-1}
9) 0.05 l
10) 2 moles
11) 166 g mol^{-1}
12) 93 g
13) 0.02 mol l^{-1}
14) 16.4 g
15) 2 l
16) 0.018 l or 18 ml

Chapter 4 Fractions

1) $0.4\ ms^{-1}$
2) $10.76\ m^2$
3) r = 1.6231699 cm V=17.9 cm^3
4) 938 cells
5) Area = 56.21 cm^2 circumference = 26.58 cm volume = 301.30 cm^3
6) $6.5\ ms^{-1}$
7) 4.5 m after 5 months 15.7 m after 9 years
8) $2.8\ cm^2\ day^{-1}$
9) $22.8\ kg\ m^{-2}$

Chapter 5 Percentage and ratio

1) 168 male
2) 3:1:1
3) 10.4 %
4) 5:3 female:male
5) 1.3 % had severe asthma, 5.9 % had mild symptoms, 92.8 % had no symptoms
6) 15.4 %
7) 0.48
8) 0.62
9) 4:9:3 robins:wrens:blackbirds
10) 1:2:4 NaCl:KOH:NaOH
11) 10% = 10 g/100 mls = 100 g l^{-1}; conc = 100/58.44 = 1.7 mol l^{-1})
12) 5% = 5 g/100 mls = 50 g l^{-1}; amount = conc. (in g l^{-1}) x vol (in litres) = 50 x 50×10^{-3} = 2.5 g
13) 24 mg ml^{-1}
14) 0.24
15) 20 times
16) 1.5×10^4 protists ml^{-1}, 0.75×10^4 protists ml^{-1}, 0.5×10^4 protists ml^{-1}.
17) 2 mg ml^{-1}
18) 1/5 dilution
19) 6.7×10^2 bacteria ml^{-1}, 16.7 %, 5.97:1
20) Cell = 12 μm
21) Organelle = 0.5 μm
22) Leg = 0.34 mm
23) Scale = 1:100,000 (0.1 m = 10 km, 1 m = 100 km = 100,000 m)
24) Distance = 1.5 km

Chapter 11 Transposing formulae

1a) $mass\,(g) = moles\,(mol) \times molar\ mass\,(g\ mol^{-1})$

1b) $mass\,(g) = molarity\,(mol\ l^{-1}) \times molar\ mass\,(mol^{-1}) \times volume\,(l)$

2) $$c = \frac{A}{\alpha l}$$

3a) $$r = \frac{C}{\pi r}$$

3b) $$r = \sqrt{\frac{A}{\pi}}$$

3c) $$r = \sqrt[3]{\frac{3V}{4\pi}}$$

4a) $$V_{\max} = \frac{V(K_m + [S])}{[S]}$$

4b) $$K_m = \left(\frac{V_{\max}[S]}{V}\right) - [S]$$

5) $$A = \frac{C^2}{4\pi}$$

Chapter 12 Graphs and their equations

1) b has the faster growth rate, a has the higher birth mass

2a)

x	y
0	-2
1	-1.5
2	-1
3	-0.5
4	0

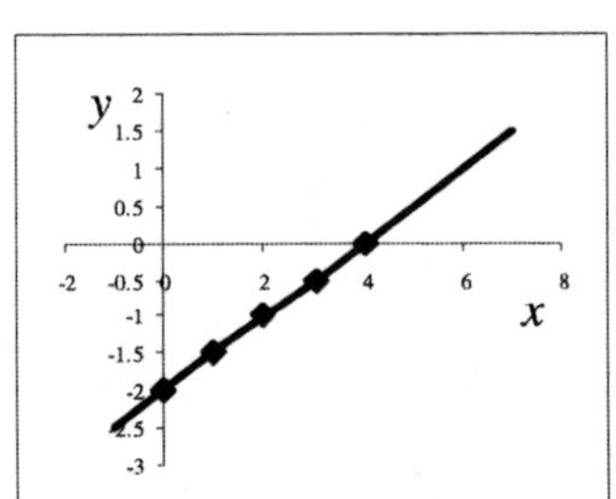

2b)

x	y
-1	3
0	1
1	-1
2	-3
3	-5

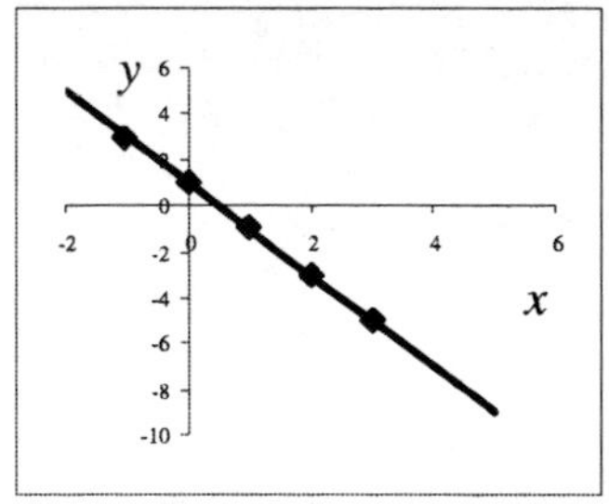

2c)

x	y
0	-1
1	2
2	5
3	8
4	11

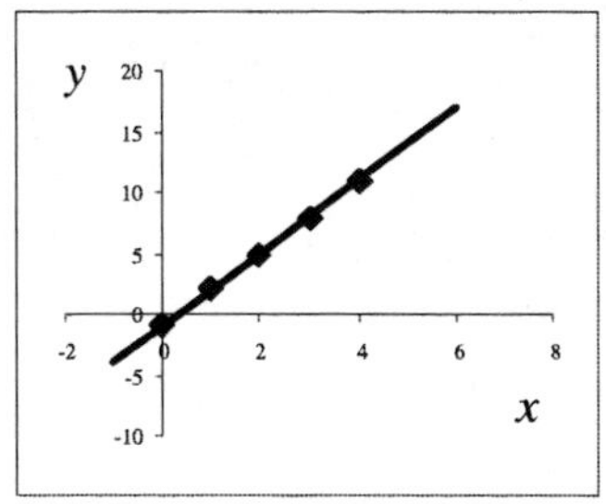

3a) y axis = A, x axis = cl

3b) $E = 3.24$, intercept, $c = 0$

4) 1.3 m min^{-1}

5)

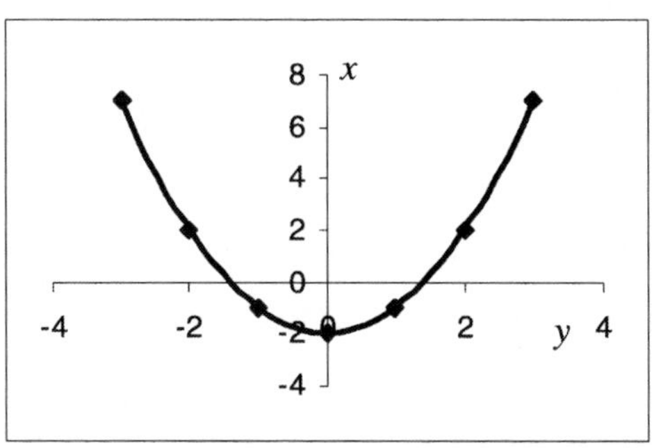

6)

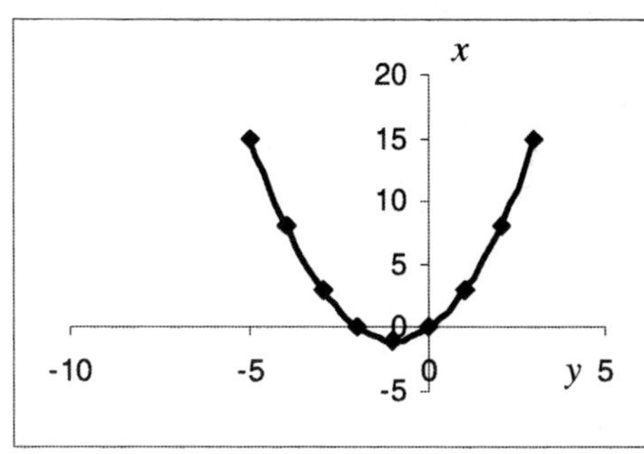

7)

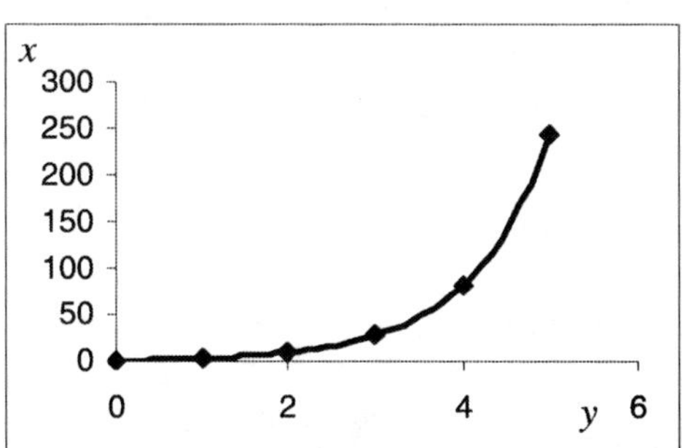

8)

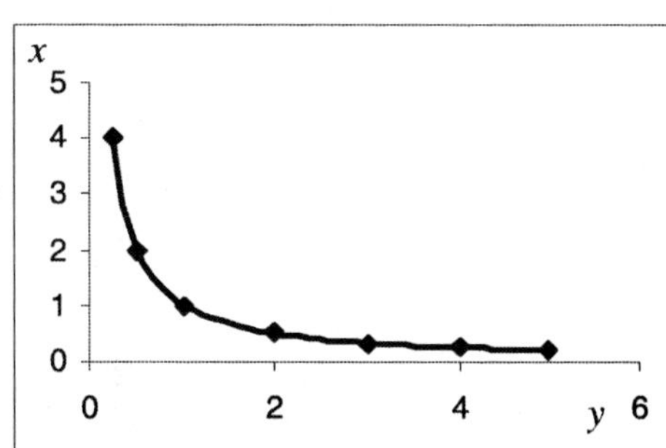

9) x would be $v/[S]$ y would be v

10) x would be $[S]$ y would be $[S]/v$

11) Gradient = k

12) Graph has a maximum point because x^2 has a minus sign in front of it

13) Use: $x = \frac{-b \pm \sqrt{b^2 - 4ac}}{2a}$ $x = \frac{-4.07 \pm \sqrt{4.07^2 - 4 \times -0.04 \times -12.70}}{2 \times -0.04}$ $x = \frac{-4.07 \pm \sqrt{16.5649 - 2.032}}{-0.08}$

$x = \frac{-4.07 \pm 3.812}{-0.08}$ $x = 3.225 \mu g l^{-1}$ $x = 98.53 \mu g l^{-1}$

14) $y = -256 + 325.6 - 12.70$ $y = 56.9 \mu m d^{-1}$

Chapter 13 Logarithm

1) $\log_2 16 = 4$

2) $\log_3 243 = 5$

3) $\log_5 25 = 2$

4) $\log_8 512 = 3$

5) $\log_4 64 = 3$

6) $2^3 = 8$

7) $9^2 = 81$

8) $3^5 = 243$

9) $4^3 = 64$

10) $7^4 = 2401$

11) 100 000

12) 4

13) 4

14) 3

15) 2

16) pH 4.4

17) pH 5.2

18) $moles\ acetic\ acid = \frac{200}{1000} \times 0.1 = 0.02 mol$

$moles\ Na\ acetate = \frac{150}{1000} \times 0.2 = 0.03 mol$

$concentration\ acetic\ acid = 0.02 \times \frac{1000}{350} = 0.57 mol l^{-1}$

$concentration\ Na\ acetate = 0.03 \times \frac{1000}{350} = 0.086 mol l^{-1}$

$pH = 4.7 + \log \frac{0.086}{0.057}$ $pH = 4.88$

19) $concentration\ of\ acid = 0.05\times\frac{1000}{300} = 0.017moll^{-1}$

$concentration\ of\ base = 0.06\times\frac{1000}{300} = 0.2moll^{-1}$ $\quad pH = 4.7 + \log\frac{0.2}{0.017} = 5.77$

20) $concentration\ of\ acid = 0.05\times\frac{1000}{2000} = 0.025moll^{-1}$

$concentration\ of\ base = 0.01\times\frac{1000}{2000} = 0.005moll^{-1}$ $\quad pH = 4.7 + \log\frac{0.005}{0.025} = 4.00$

21) Formic acid HCOOH splits into H^+ and its conjugate base $HCOO^-$ $\quad HCOOH \rightleftharpoons H^+ + HCOO^-$

Initially the concentration of HCOOH =*A*, and the concentration of H^+ and $HCOO^-$ = 0. At equilibrium the concentration of HCOOH = *A-X* and the concentration of both H^+ and $HCOO^-$ = *X*

K_a (dissociation constant) for the acid is given by equation:

$$K_a = \frac{X^2}{A-X}$$

where $K_a = 1.74\text{x}10^{-4}$
X is the concentration of H^+ ions
A is the starting concentration of formic acid

First you need to rearrange the formula to produce a quadratic equation that can be solved for *X*, the hydrogen ion concentration.

$K_a(A-X) = X^2 \qquad X^2 + K_aX - K_aA = 0$ $\quad$ this is in the form $\quad ax^2 + bx + c = 0$

so use the equation to solve a quadratic $\quad X = \frac{-K_a \pm \sqrt{K_a^{\ 2} + 4K_aA}}{2}$

$$X = \frac{-1.7\times10^{-4} \pm \sqrt{(1.7\times10^{-4})^2 + 4\times1.7\times10^{-4}\times0.02}}{2} \qquad X = \frac{-1.7\times10^{-4} \pm 3.68\times10^{-3}}{2}$$

There are two solutions, but one is impossible, so $X = 1.79\text{x}10^{-3}$. Now you use $\quad pH = -\log[H^+]$

$pH = -\log[1.79\times10^{-3}] \qquad pH = 2.75$

22) $8.0 = 8.08 + \log\frac{base}{acid} \qquad \frac{base}{acid} = 0.83$ $\qquad$ [acid]:[base] = 1.2:1

23) pH = 5

24) pH = 8

25) pH = 3.3

26) pH = 11

27) pH = 12

28) pH 6.3

29) 16.5 years

30) 0.0819 y^{-1}

31) 1.98 days

32) 0.023 s^{-1}

Chapter 14 Calculus

1) $\frac{dy}{dx} = 2x + 3$

2) $\frac{dy}{dx} = 4x - 5$

3) $\frac{dy}{dx} = 3x^2 + 4x$

4) $\frac{dy}{dx} = -6x^{-3}$

5) $\frac{dy}{dx} = 12x^2$

6) $\frac{dy}{dx} = -12x^{-4}$

7) $\frac{dy}{dx} = 2x - 4$

8) $\frac{dy}{dx} = 3x^2 + 3$

9) $\frac{dy}{dx} = 8x - 3$

10) $\frac{dy}{dx} = 1$

11)

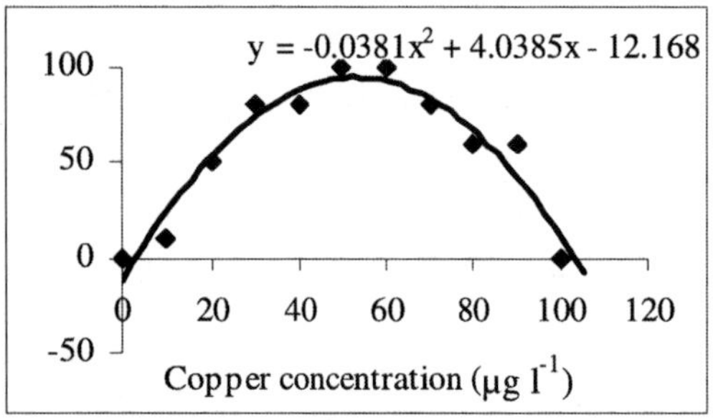

$$\frac{dy}{dx} = -0.0762x + 4.0385$$

$$-4.0385 = -0.0762x$$

$$52.998 = x$$

Copper concentration = 53 µg l^{-1}

$$y = -0.0381 \times 53^2 + 4.0385 \times 53 - 12.168$$

$$y = -107 + 214 - 12.168$$

$$y = 94.8$$

Maximum growth = 94.8 µm d^{-1}

12) A tangent to a curve is a line that touches the curve i.e. it intersects it once only. A secant line of a curve is a line that intersects it at two or more points.

13) $\frac{\Delta y}{\Delta x}$ is the average rate of change of y over the interval Δx

$\frac{\delta y}{\delta x}$ is the average rate of change of y over a small interval δx

$\frac{dy}{dx}$ gives the instantaneous rate of change of y for any given value of x.

14) $\frac{dy}{dx} = 3x^2$

15) $\frac{dy}{dx} = -\frac{2}{x^3} = -2x^{-3}$

16) $\frac{dy}{dx} = \frac{1}{x^2} = x^{-2}$

17) $\frac{dy}{dx} = 3x^2 + 2x$

18) $\frac{dy}{dx} = 4x^3 - 4x - \frac{6}{x^4}$

19) $\frac{dy}{dx} = -\frac{3}{x^4} + \frac{2}{x^2}$

20) $\frac{dy}{dx} = \frac{1}{2}x + 2x = \frac{5}{2}x$

INDEX